Biochemical Disorders
of the Skeleton

POSTGRADUATE ORTHOPAEDICS SERIES
under the General Editorship of
A. Graham Apley, MB, BS, FRCS
Honorary Director, Department of Orthopaedics, St. Thomas' Hospital, London;
Consultant Orthopaedic Surgeon, The Rowley Bristow Orthopaedic Hospital, Pyrford
and St. Peter's Hospital, Chertsey, Surrey

Biochemical Disorders of the Skeleton

Roger Smith

MD, PhD, FRCP
Consultant Physician, The Radcliffe
Infirmary and The Nuffield Orthopaedic
Centre, Oxford

Butterworths
LONDON-BOSTON
Sydney-Wellington-Durban-Toronto

The Butterworth Group

United Kingdom	**Butterworth & Co (Publishers) Ltd**
London	88 Kingsway, WC2B 6AB
Australia	**Butterworths Pty Ltd**
Sydney	586 Pacific Highway, Chatswood, NSW 2067
	Also at Melbourne, Brisbane, Adelaide and Perth
South Africa	**Butterworth & Co (South Africa) (Pty) Ltd**
Durban	152–154 Gale Street
New Zealand	**Butterworths of New Zealand Ltd**
Wellington	T & W Young Building, 77–85 Customhouse Quay, 1, CPO Box 472
Canada	**Butterworth & Co (Canada) Ltd**
Toronto	2265 Midland Avenue, Scarborough, Ontario, M1P 4S1
USA	**Butterworth (Publishers) Inc**
Boston	19 Cummings Park, Woburn, Mass. 01801

First published 1979

ISBN 0 407 00122 0

© Butterworth & Co (Publishers) Ltd 1979

British Library Cataloguing in Publication Data

Smith, Roger
  Biochemical disorders of the skeleton. — (Postgraduate orthopaedics series).
  1. Musculoskeletal system — Diseases
  2. Physiological system
  I. Title II. Series
  616.7'1'071   RC925   78-41076

ISBN 0 407 00122 0

Typeset by Butterworths Litho Preparation Department
Printed and bound at William Clowes & Sons Limited, Beccles and London

Editor's Foreword

Every surgical trainee today learns some biochemistry. Unhappily, by the time he has been in consultant practice for a few years, most of it is forgotten. Surgeons of my own vintage are perhaps more fortunate; we had almost nothing to forget. Yet, old or young, all of us need to know a little or we fail to understand so many disorders; and the more we know the more we want to know, because metabolic disorders have the fascination of a chess problem.

Nor is such learning mere formula-hunting (the unteachable pursuing the unknowable?). Admittedly some of the diseases are distinctly uncommon, so that we tend to think of the whole group as a collection of rarities. Far from it. Some, like osteoporosis and Paget's disease, are among the commonest bone affections of mankind; even the rarities add up to a considerable total, and all have biochemical anomalies.

When first I heard Roger Smith on metabolic disorders I was delighted. Here was a physician whose grasp of their scientific basis was such that he could present the biochemical facts with simplicity and with clarity. And yet at the same time his clinical perceptiveness totally rivetted my attention. A book on the subject was needed, so what more natural than to ask him to write it? I believe physicians and paediatricians may get just as much from this book as will orthopaedic surgeons. I hope they will enjoy learning from it (or rather looking something up and then reading on compulsively, as we do with Fowler) as much as I have enjoyed editing it.

A.G.A.

Contents

Preface

The mysteries of the skeleton continue to attract people from many disciplines and their work contributes to the increasingly rapid advances in this area of medicine.

Many of these advances are biochemical, but they contain clear clinical messages and an understanding of them can be of direct benefit to the patient.

However it is natural that there are many gaps in communication between those who see most of the patients (the orthopaedic surgeons), those physicians who are interested in bone disease, and those who work in the laboratories.

In this short book I have tried to bridge some of these gaps and to bring together the biochemical and clinical aspects of disorders of the skeleton.

The scope of this book is wider than that of the classic metabolic bone diseases with their emphasis on the mineral phase of bone, since it also includes clinically important conditions arising from disorders of the organic bone matrix.

Within its size the book cannot be comprehensive, and it is certainly not intended as a text on 'metabolic bone disease', of which several now exist. What it does try to do is to provide a description of those biochemical disorders which primarily or significantly affect the skeleton and which are met with both in orthopaedic and medical practice; and, where possible, to explain the clinical features in terms of the biochemical defect.

I hope that within this book orthopaedic surgeons will find help in diseases made needlessly complex by physicians; and that physicians will find points of interest in disorders which they have previously regarded as exclusively orthopaedic. Further, since these pages are partly based on postgraduate lectures, they may be of some general interest.

With regard to the construction of the book, Chapters 1 and 2 deal with physiology and diagnosis, and it would be logical to read them first. Apart from this the chapters can be read in any order and are written to be largely complete in themselves; to this end there is some repetition of the physiology in appropriate chapters. The references at the end of each chapter have been chosen either to show work which is classic or important and recent, or to support the statements made. Some attempt has also been made to give references to the most easily accessible journals.

There are two other aids. One is a glossary of abbreviations, and the other is a short description of recent advances (Chapter 12), put into the text at the latest possible moment in an attempt to keep the reader up to date.

Finally, should the reader require more information, references are given in Chapter 1 and elsewhere to suitable larger text books. Whilst there is some choice in metabolic bone disease there is no acceptable alternative to McKusick's book, *Heritable Disorders of Connective Tissue*. Unfortunately, all books (including this one) are ageing by the time they are published, and the real enthusiast should scan the journals for review articles, particularly in the *New England Journal of Medicine*.

Much of this book has been influenced by the work of my Oxford colleagues, Martin Francis, John Kanis, Graham Russell, Bryan Sykes and Colin Woods, to whom I am very grateful. I know that they will agree with me that any merit is theirs and that any mistakes are mine.

R.S.

Acknowledgements

I am particularly indebted to Hilda Moore for her patience in typing the manuscript and to Bob Emanuel and his staff for photographing the diagrams and radiographs. Some of these diagrams were kindly drawn by Graham Bartlett and Peter Brooks.

I am very grateful to Dr Colin Woods for providing *Figures 1.2, 3.5, 4.1* and *5.1* and to Dr David Frazer for *Figure 1.14.* Mr Tony Bron kindly provided the slides of abnormalities of the eyes from which *Figures 2.5* and *8.5* were prepared by the Photographic Department of the Radcliffe Infirmary.

Several authors, journals and publishers have kindly given their permission for use of the following figures:

Figures 1.5	: Drs Grant and Prockop, *New England Journal of Medicine*
1.6	: Dr Pinnell, McGraw-Hill
1.8	: Dr Francis, *Clinical Science,* Blackwell Scientific Publs
1.9	: Dr Sykes, *Biochemical and Biophysical Research Communications,* Academic Press
1.10	: Dr Herring, *Clinical Orthopaedics,* J. B. Lipincott Co.
1.11	: Dr Rasmussen, Williams and Wilkins
1.13	: Dr Janet Vaughan, Clarendon Press
1.17	: D. H. Copp, *Journal of Endocrinology*
2.1. 2.2	: Dr Round, *British Medical Journal*
2.7	: Dr Walton, *Clinical Science and Molecular Medicine,* Blackwell Scientific Publs
2.8	: Prof. Nordin, Churchill Livingstone
2.9	: Dr S. Clark, Academic Press
3.1	: Dr Morgan, *Clinical Orthopaedics,* J. B. Lipincott Co.
3.2	: Dr Smith, *Journal of Clinical Investigation,* Rockefeller University Press
4.11	: Dr Rasmussen, *New England Journal of Medicine*
5.10, 5.11	: Dr Walton, *European Journal of Clinical Investigation,* Blackwell Scientific Publs

5.13   : Dr Whiteley, *Clinica Chimica Acta,* Elsevier/N. Holland
            Biomedical Press
6.3   : Dr Byers, *Quarterly Journal of Medicine,* Oxford University
            Press
6.5   : Dr Peacock, Churchill Livingstone
6.6   : *Clinica Chimica Acta,* Elsevier/N. Holland Biomedical
            Press
6.7   : Dr Tomlinson, *Hospital Medicine,* Northwood Publs
6.10   : Dr Russell, *Lancet*
7.1   : Dr Bauze, *Journal of Bone and Joint Surgery,* E. & S.
            Livingstone
7.5   : Dr Levin, The National Foundation — March of Dimes
7.12   : Dr Sykes, *New England Journal of Medicine*
9.1   : Dr McKusick, Pitman Medical
10.5, 10.7 : *Seminars in Arthritis and Rheumatism,* Grune and Stratton
Tables 1.II   : Dr Ashton, *Calcified Tissue Research,* Springer Verlag
        5.I, 5.II : Dr Walton, *Calcified Tissue Research,* Springer Verlag

ERRATA

Page 52, final paragraph, line 1 should read:
Although the plasma calcium, phosphate, and alkaline phosphatase are the main . . . .

Page 53, final paragraph, line 1 should read:
The relationship between calcium excretion (Ca_E) per 100 ml glomerular . . . .

Page 140, line 1 of text should read:
of the ear. The optic nerve is virtually never affected. Diplopia, dysphagia . . . .

Page 164, line 2 should read:
(7 per cent), hypertension (4 per cent), association with other endocrine
disorders . . . .

Page 277, line 21 should read:
adenyl cyclase enzyme may be abnormal. Treatment with $1,25(OH)_2D$ is . . . .

1

Physiology of Bone and its Control

INTRODUCTION

We tend to take the skeleton for granted. During most or all of our lives it performs its structural function without attention or fault, and in comparison with other organs, such as the heart or lungs, its occasional failure does not threaten life.

There are those who regard bone as a sort of human reinforced concrete, in which the organic matrix provides the steel and mineral the concrete. Superficially this view is understandable since the skeleton is of considerable strength, gives every outward sign of inertness, and persists long after death.

However, if we look beyond the obvious, we soon realize that bone is an active tissue which undergoes complex structural and metabolic changes. It is a tissue apparently enclosed in its own extracellular fluid in which specialized cells under hormonal control constantly produce and destroy both matrix and mineral. It is the presence of this mineral which gives the impression of permanence.

It is our task in this introductory chapter to consider the cells of bone, the structures they produce, and the many hormones which influence them.

Function

Bone is considered to have two main functions, the mechanical support and protection of the body, and the maintenance of normal mineral metabolism. The control of these respective functions is often referred to as skeletal and mineral homeostasis (Vaughan, 1975). This is a convenient division and one which serves to separate subjects about which we know little, such as the control of skeletal growth and its maintenance by mechanical stress, from those about which we think we know more, such as the control of calcium metabolism. However, a little thought will tell us that this separation is artificial. For example, growth

1

and development of bone must affect mineral metabolism, as will rapid changes in the adult skeleton such as those due to immobilization; and overactivity of endocrine glands such as the parathyroids will disturb both biochemistry and structure.

Recent advances

Bone is a tissue which is difficult to study directly. The information about its composition and physiology is therefore still incomplete and controversial, and new ideas are often based on inference rather than fact. Despite this, at least some of the recent work is likely to be correct and deserves our attention.

One important idea, which was developed by Neuman (1969) and his colleagues, and for which there is increasing evidence, is that the bone surfaces are bathed in a bone fluid which differs markedly in composition from the whole body extracellular fluid, from which it is separated by a sheet of cells including osteocytes and osteoblasts. More controversial is the work concerning the origins and functions of such cells (p. 4). Thus the established view of the separate origins of osteoblasts and osteoclasts has been challenged by Rasmussen and Bordier (1973, 1974), who postulate that osteoblasts are in fact derived from osteoclasts. Although the evidence for this is not convincing, if true it would neatly explain the known linkage between the rates of bone resorption and formation. Certainly the osteocytes of bone may have a far wider function than had previously been supposed; and they may be rightly regarded as an important component of the bone cell system for mineral and skeletal homeostasis, rather than as obsolete osteoblasts incarcerated in their own mineral.

In the chemistry of bone, the importance of the organic matrix is being increasingly recognized, although little is yet known of the factors which control its metabolism. However, it is now realized that collagen exists in different genetic types and that disorders may arise from defects in its synthesis, post-translational modifications, and cross-linkage. Additionally, relevant components of the minor non-collagen protein fraction of bone have been identified, which include a specific glycoprotein of plasma apparently made by the liver and incorporated into mineralizing bone.

With regard to mineral homeostasis, the reader will be aware of the rapid advances which have been made in knowledge of the metabolism and actions of vitamin D, and will be justifiably confused by the mass of literature such advances have produced. Less dramatic but important progress has also been made in understanding the action of parathyroid hormone and in demonstrating the effects of prostaglandins and related substances on the skeleton. Many fundamental mysteries still remain; for example, we do not yet know the physiological role of calcitonin; we have no idea how stress on bone stimulates its formation (despite the interesting studies on piezo-electricity (Bassett, 1971)); and the mechanism of mineralization, a process which distinguishes bone from all other tissues, remains unknown.

Blood supply and growth

The supply of blood to the bone and its normal growth are subjects which are not dealt with in this book. This does not imply that they are of no significance, since the opposite is the case. The anatomy of the vascular supply to the bone is well established (Brookes, 1971), but there are considerable difficulties in measuring the rate of flow within it. Wootton, Reeve and Veall (1976) emphasize that the metabolism of any tissue is partly regulated by its blood supply and that this is presumably true for bone. They have developed clinical methods for the measurement of blood flow which depend on the complete extraction of ^{18}F in a single passage through bone, have established normal values, and demonstrated considerable increases in Paget's disease of bone (Chapter 5).

The normal growth of the skeleton, like that of any other tissue, is an astonishing process about which we know very little. If we merely consider the changes in the growing ends of a long bone there may be many interrelated processes occurring simultaneously, upset of any of which could result in a catastrophic disorder. For example, the transformation from cartilage to bone involves a change in molecular species of collagen, alteration of proteoglycan content, mineralization aided by extracellular matrix vesicles, and continual cellular activity to model and remodel the bone into its final shape. These processes are presumably controlled both by hormones, and by mechanical forces, but our knowledge of how they act is fragmentary. For reviews of the growth of bone, the reader should refer to Bourne (1972) and Vaughan (1975).

COMPOSITION OF BONE

The main constituents of bone are the bone cells, the organic bone matrix* and the mineral which is laid down upon it. Other components include nerves, blood vessels and marrow. Bone, like other tissues of the body, is metabolically active and its formation, composition and resorption are primarily controlled by cellular events, which are therefore of first importance.

We should enquire where the bone cells come from, what they do, and why. In the account which follows the histology refers to adult bone and most often to its endosteal surface. Much information comes from the extensive review of Parfitt (1976), which the reader should consult, and from Rasmussen and Bordier (1973, 1974), Vaughan (1975), and Jowsey (1977). Current as well as more established views are expressed. Aaron (1976) gives a good review of the micro-anatomy of bone and Owen (1977) some useful references. The mineral aspects of bone physiology are well described in various chapters of the book by Nordin (1976), but for those of matrix the reader should look elsewhere (e.g. Hall and Jackson, 1976).

Bone cells

The bone cells which mainly concern us are the osteoblasts, the osteoclasts and the osteocytes. Fibroblasts also are found in the periosteum, the marrow and around blood vessels. The marrow also contains the full series of myeloproliferative cells, including undifferentiated stem cells.

---

*By common usage the term matrix is applied to the organic phase of bone, in contrast to the mineral. Matrix may also be used as a term to distinguish the structure of formed bone from the soft tissues, and the cells which circulate within it.

Origin of bone cells

The source of bone cells has been much debated. Available evidence (Owen, 1977) continues to support the view that osteoblasts and osteoclasts have separate lines of cellular differentiation (*Figure 1.1*). Bone marrow contains two histogenetically distinct cell lines, the haemopoietic and the reticulo-endothelial systems. There is good evidence that the precursors of the osteoclasts are monocytic cells of the haemopoietic system, and that the reticular (stromal) cells are the precursors of

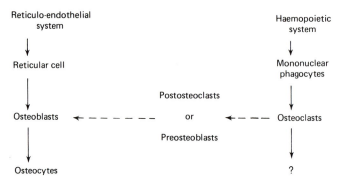

Figure 1.1. To demonstrate the separate origin of osteoblasts and osteoclasts in the postnatal organism. The interrupted line indicates the Rasmussen and Bordier (1973, 1974) postulate that osteoclasts are precursors of osteoblasts. Some osteoblasts are incorporated as osteocytes into mineralizing bone. The fate of osteoclasts is not clear

osteoblasts. Thus in transplantation experiments bone is formed from the marrow stromal cells of the donor, whereas the haemopoietic tissue comes from the recipient. Contrary to these findings are the proposals of Rasmussen and Bordier (1973, 1974), that at least on the endosteal surface of bone, the precursors of osteoblasts are osteoclasts.

Irrespective of the origin of bone cells, it appears that remodelling of established bone occurs in units (Frost, 1964), in which resorption precedes formation (p. 9). Thus activation of a new group of cells is said to take place normally over a period of a few hours or days; the resorptive (osteoclastic) phase lasts one to three weeks, and the final phase of bone formation may continue for more than three months. One practical result of this time sequence (which is not always remembered by clinicians) is that treatment will not produce rapid changes in bone structure detectable by routine bone histology, although it may quickly cause selective changes in the number and appearance of bone cells.

Since intermittent remodelling occurs throughout the skeleton with new units being activated and with cycles being completed, it is reasonable to enquire what happens to the osteoblasts after the end of bone formation, and it is likely that they become osteocytes within the mineralized bone. However Rasmussen and Bordier (1973) suggest that such osteocytes may recapitulate the sequence of cellular events characteristic of the bone remodelling unit, and they find it possible to describe the theoretical effects of hormones and ions on the modulation and activity of bone cells, interpreting the changes in disease according to these cycles of activity.

TABLE 1.I
The bone cells and their functions

	Appearance	*Function*
Osteoblast	Plump, uniform in size. Regular array on underlying bone. Processes extend into osteoid	Formation of bone matrix. Involved in calcification
	High content alkaline phosphatase	
Osteoclast (*Figure 1.2*)	Large. Multinucleated,* mobile. Contain lysosomes and many mitochondria. Ruffled border extends into mineralized bone. High content acid phosphatase	Bone resorption
Osteocyte	Variable, according to supposed function. In young woven bone – numerous short processes, fill lacunae. In adult lamellar bone – flattened, long processes, do not fill lacunae	Debated; the osteocyte-canalicular system is important for rapid mineral exchange

* Occasionally osteoclasts may appear to be mononuclear

The continuing controversy on the origins of bone cells need not further concern us here. Since, however, bone cells are as essential to the skeleton as other cells are to any metabolically active tissue the reader should know what the bone cells look like and particularly what they are supposed to do (Table 1.I).

The osteoblast

(a) Structure
Osteoblasts, derived from recognizable precursors, preosteoblasts, are the bone forming cells. In appearance they are plump with basophilic cytoplasmic and electron microscopic characteristics which indicate active protein synthesis. They are found in single layers, closely applied to the unmineralized bone matrix (osteoid) of newly forming bone. They appear to form a barrier between the surface of the forming bone on one side and the connective tissue and blood vessels of the marrow on the other. Some areas of the cells make contact with each other by 'tight' or 'gap' junctions (*see* Parfitt, 1976), but in many areas channels appear to run between the cells. The osteoid seams, upon which the osteoblasts lie, are separated from fully mineralized bone by the mineralization or calcification front. This front has characteristic staining properties and takes up tetracyclines. Electron microscopy shows that fine protoplasmic processes from the osteoblasts penetrate the osteoid through to the mineralized bone, and finally rest on the plasma membranes of deeper osteocytes or their processes. Extensions from the osteoblasts within the osteoid may possibly form the beginnings of calcium-accumulating vesicles (p. 20).

It is usual to divide osteoblasts into 'resting' and 'active', the former being spindle-shaped and flattened, the latter being big and plumper. Rasmussen and

Bordier (1974) regard resting osteoblasts as surface osteocytes which are in contact with the physiologically important system of osteocytes within the canaliculi of bone.

(b) Function

The osteoblast has many functions. One of the most important is the synthesis of collagen (p. 12), the main constituent of bone matrix. It also manufactures the 'ground substance' which contains the glycoproteins of bone. What is not clear is how the osteoblast contributes to the mineralization of the matrix which it has produced.

Bone appears to be made by osteoblasts in successive stages of matrix formation and mineralization. The formation of matrix involves synthesis of collagen and non-collagen protein (proteoglycan and glycoproteins). Mineral is thought to be originally deposited as mainly amorphous tricalcium phosphate, which is slowly converted to crystals of hydroxyapatite. The process is said to occur in two stages; up to 75 per cent of the mineral appears to be deposited within a few days and is under the control of osteoblasts; continuing mineralization is not under osteoblastic control and goes on for several months.

The osteoblast, like other cells, contains calcium and phosphate and some of the calcium is particularly associated with the mitochondria. Further, the calcium-accumulating vesicles said to be present in the intercellular matrix of bone (as well as in mineralizing cartilage) may be derived from the osteoblasts, and certainly appear to have a cellular origin. They contain numerous enzymes including pyrophosphatase and ATPase. The osteoblast itself is characterized by its extremely high content of alkaline phosphatase (which can act as a pyrophosphatase), and it is difficult to escape the conclusion that the osteoblast and extensions derived from it are importantly concerned with mineralization (*see* Vaughan, 1975; and Jowsey, 1977, Figure 9-4).

The osteoclast

(a) Structure

Osteoclasts, derived from haemopoietic precursors, are to be found on resorbing surfaces, and are characteristically multinucleated and large. Because of their resorptive activity they come to lie in small depressions called Howship's lacunae. Apart from voluntary muscle fibres they are the largest cells in the body (Aaron, 1976). Their cytoplasm is full of vacuoles which resemble lysosomes and presumably contain the enzymes responsible for the digestion of bone. The brush or ruffled border of the osteoclast is applied to the bone surface with the cytoplasmic process extending from the osteoclast to the bone. Between these processes the initial steps of digestion appear to occur (*Figure 1.2*).

(b) Function

The osteoclast is a mobile bone-eating cell. Its activity appears to correlate with the presence and extent of the ruffled border which infiltrates the disintegrating bone surface. The exact mechanisms of bone resorption are debated, but the cell has a full complement of enzymes which are capable of working in acid or neutral environments. It may be that mineral is removed before matrix but for practical purposes they are removed together. Removal of the mineral phase has

6

Figure 1.2. The electron microscopic appearance of an osteoclast on a bone surface. The ruffled border of the osteoclast (right) is applied to the bone surface (left). Fragments of collagen are being digested. Magnification × 18 000

been attributed to the local production of citrate and lactic acid. The matrix is presumably digested by enzymes but controversy exists about which are responsible. After primary cleavage of the collagen fibril by collagenases, further breakdown may occur in the lysosomes of the osteoclasts (p. 15). Lysosomal enzymes also degrade proteoglycans.

For unknown reasons osteoclastic resorption of osteoid is rarely seen. The presence of an osteoclast is the only means we have of distinguishing a surface undergoing resorption at the time of sampling, from one where resorption has ceased.

The osteocyte

(a) Structure

This cell may be regarded as an osteoblast which has become imprisoned in mineralized bone. This derivation from the osteoblast has been shown, but there is much argument about its function within the bone. The cells vary much in shape, from the early osteocytes which resemble osteoblasts, and may be found with them on bone surfaces, to the more mature flattened osteocytes which appear not to fill the lacunae in which they lie and communicate with each other through fine branching processes which extend into the canaliculi of the mineralized bone. The manner in which osteocytes probably communicate with others within bone and on its surface is suggested in *Figure 1.3*. The osteocyte is the single most numerous type of bone cell.

Aaron (1976) points out that the highly variable histological appearance of the osteocyte is compatible with its altering function. She also states that the importance of the osteocytes becomes clear when they die, because the surrounding matrix becomes structurally inadequate and is removed to be replaced by living bone. However, since it is not possible to selectively destroy osteocytes the maintenance of normal bone structure may depend on other bone cells as well.

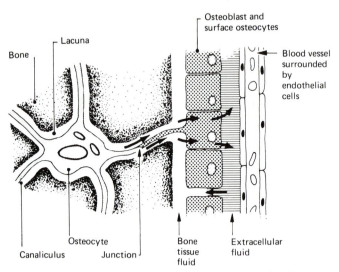

Figure 1.3. Osteocytes within lacunae communicate with other osteocytes within bone and with osteocytes on the bone surface. Ionic transfer may occur within the cytoplasmic processes and via the bone fluid in the lacunae and canaliculi. The nature of the junctions between the protoplasmic processes is not known, and the direction of flow in the bone fluid is not decided (see Parfitt, 1976). The bone surface is covered by a layer of cells separating bone tissue fluid from extracellular fluid. For clarity the osteoid layer covering the newly formed bone and between it and the bone tissue fluid is omitted

It is important here to re-emphasize that bone surfaces are covered by a layer of cells placed between them and the capillaries. Thus there is at least over the majority of bone surfaces a complex controlling mechanism through which metabolites must pass to reach bone. Within bone, these metabolites may be transported outside the osteocytes, i.e. in the extracellular fluid of bone (the bone fluid) within the canaliculi, or within cells and from one cell to another.

Parfitt (1976) deals with this point in more detail. He also emphasizes the peculiarity of bone adjacent to the wall of the osteocyte (perilacunar bone) compared with bone elsewhere. The collagen fibres are fewer and less densely mineralized and the mineral is more soluble and amorphous. Perilacunar bone has different staining properties from other bone, and is also considered to be of particular physiological importance in calcium homeostasis. Although the lacunar–canalicular system is well described and has a considerable surface area for metabolic exchanges, a much larger surface area for exchange between crystal surface and bone fluid is provided by microcanaliculi radiating from the canaliculi into the adjacent bundles of collagen fibrils (*see again* Parfitt, 1976).

(b) Function

The role of the osteocyte is not established. Possible functions attributed to it include those of metabolic activity, of osteolysis, as a bone 'pump', as a system concerned with long-term calcium exchange and also with plasma protein uptake (Vaughan, 1975). Osteocytes probably do different tasks in different places. Thus surface osteocytes may be more active than those within mineralized bone; but it would be wrong to regard the latter as inert. Osteocytes may synthesize at least some bone collagen and control its mineralization. However it is when we come to consider whether they also cause bone resorption or removal of mineral without matrix that opinions differ. This is because, in addition to the usual difficulty of interpreting histological appearances in terms of function (Jowsey, 1977), the normal appearance of osteocytes within their lacunae alters as the bones age. Some bone pathologists regard the appearance of small osteocytes within large lacunae as evidence of active osteolysis by the osteocytes, particularly in hyperparathyroidism. Boyde (1972) using the scanning electron microscope found that in rats given parathyroid hormone there was no evidence of osteocytic osteolysis. These difficulties in deciding what the osteocyte does from its appearance are also true for other bone cells. However since the activity of cells is fundamental for the maintenance of the normal skeleton, any method of examining bone, such as microradiography, which does not demonstrate them, must be incomplete.

Modelling and remodelling

The cells of bone carry out two important and related structural processes. These are the modelling of bone during growth and the remodelling of adult bone after growth has ceased. These processes are complex. The following short account is taken from the admirable review of Parfitt (1976) and deals with remodelling first.

Remodelling has been studied best in cortical bone. Since this bone is dense, new bone can be made only if old bone is removed. This process of removal is carried out by a well defined cutting cone of osteoclasts advancing longitudinally through the cortex. Formation is performed by a corresponding closing cone of osteoblasts. Studies on these mechanisms, mainly carried out in the dog, provide evidence of an anatomically distinct structure which persists for a variable time and is responsible for bone turnover in an ordered and predictable manner. Such studies cannot be directly applied to trabecular bone in man, although analogous appearances are seen, but they provide much fascinating information for future work. However the nature of coupling between resorption and formation still remains completely unknown.

In contrast in bone modelling during growth, resorption and formation are segregated to different surfaces and continue at the same surface without interruption for various lengths of time. Growth in length occurs by endochondral ossification, and in width by periosteal apposition and endosteal resorption. At the growing end of a long bone chondrocytes are arranged in columns parallel to the long axis of the bone. These chondrocytes go through successive stages of development as the epiphyseal plate moves away from them, becoming rounder and larger, and passing through the so-called proliferative, hypertrophic and provisional calcification zones. Calcification occurs in the longitudinal septa between the columns of cells (most of which degenerate), spaces between the calcified bars are invaded by capillaries, the transverse septa between the cartilage cells are removed, and primitive mesenchymal cells from the diaphyseal region differentiate into chondroclasts (which partly resorb the calcified longitudinal septa) and osteoblasts (which lay down woven bone on their surfaces). This combination of partly removed calcified cartilage encased in woven bone may be referred to as the primary spongiosa. The secondary spongiosa develops from this by piecemeal osteoclastic resorption, followed by osteoblastic and haemopoietic stem cells.

The dissociated events of endosteal resorption and periosteal apposition in the shaft of a growing bone may be modified by subperiosteal resorption to shape the metaphyseal regions, and also by eccentric growth. Defects in modelling are well seen in some of the osteopetroses (Chapter 11) where osteoclastic resorption is defective.

The role of the osteocyte in these processes is again debatable. Parfitt (1976) is of the opinion that periosteocyte remodelling is probably of small magnitude. He further emphasizes the confusion produced by the widespread use of the term osteocytic osteolysis, which he considers is best applied to pathological enlargement of the lacunae (which may become confluent) in response to an abnormal resorptive stimulus. The significance of apparent resorption of bone around osteocytes and without osteoclasts has yet to be defined.

Bone matrix

The principal constituent of bone matrix is collagen, but the matrix also contains a complex mixture of compounds containing protein and carbohydrates. These are variously named as glycoproteins or proteoglycans (protein-polysaccharides) according to their structure. Their functions are largely unknown.

Collagen

Collagen is the main extracellular protein of the body, it is metabolically active, and more than half of it is in bone (Grant and Prockop, 1972). These important facts have been neglected by those interested in metabolic bone disease since diagnosis and investigation has been predominantly directed to the mineral components of bone*. Of the diseases considered in this book there are some where a chemical abnormality of collagen provides the main clinical features (Chapters 7 and 8), but in addition collagen is involved in all those disorders which affect mineralized bone, such as hyperparathyroidism, Paget's disease and osteomalacia. Further, some of those hormones thought primarily to affect mineral metabolism, such as vitamin D, may also have an effect on bone matrix. It is important to measure the biochemical changes produced by alterations in bone matrix, but until recently such measurements were difficult. This section will consider briefly the synthesis, structure and breakdown of collagen and the

---

*Avioli and Krane (1977) point out that in Albright and Reifenstein's book (1948) collagen was never mentioned

ways in which it can be examined. Good reviews are given by Grant and Prockop (1972); Nimni (1974); Bornstein (1974); Harris and Krane (1974) and Fietzek and Kuhn (1976).

(a) Structure

The molecule of collagen (also called tropocollagen) is made from 3 polypeptide (α) chains wound into a triple helix. Each chain which contains approximately 1000 amino acids is initially synthesized in the same way as other proteins and subsequently modified. It used to be thought that there was only one type of collagen composed of two α_1 chains and one α_2 chain, but it is now realized that there are at least four genetically different collagens, namely Type I $[\alpha_1(I)]_2\alpha_2$, Type II $[\alpha_1(II)]_3$, Type III $[\alpha_1(III)]_3$, and Type IV $[\alpha_1(IV)]_3$, which occur in different tissues (*Figure 1.4*) and have biochemical differences. Thus Type I collagen, which occurs alone in bone, has a low amount of carbohydrate and less

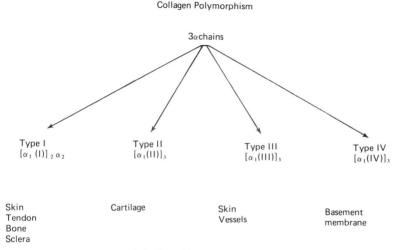

Collagen Polymorphism

3αchains

Type I	Type II	Type III	Type IV
$[\alpha_1(I)]_2\alpha_2$	$[\alpha_1(II)]_3$	$[\alpha_1(III)]_3$	$[\alpha_1(IV)]_3$
Skin	Cartilage	Skin	Basement
Tendon		Vessels	membrane
Bone			
Sclera			

Figure 1.4. The different genetic collagens

than 10 hydroxylysine residues per chain: Type II collagen has more than 10 hydroxylysine residues per chain and about 10 per cent carbohydrate, and Type III collagen contains cysteine, a larger amount of 4-hydroxyproline and glycine, and little carbohydrate. Basement membrane collagen is high in 3-hydroxyproline. Only Type I collagen contains different genetic types of collagen chain (i.e. is a heteropolymer).

Collagen fibres are composed of cross-linked collagen molecules arranged in a quarter stagger array. The molecules themselves are about 3000Å* long, and the arrangement produces an electromicroscopic appearance with a 680Å banding pattern, shown as D in *Figure 1.5*. This diagram also shows the amino acid sequence of the helical position of the chains, and the arrangement of the chains themselves. It is possible to arrange the collagen molecules in register without overlapping, as so-called segment long-spacing (SLS) crystallites, when electron

*Now expressed as nanometres. 10Å = 1nm

microscopy reveals many details of the molecules not seen in the native fibre (*see* Fietzek and Kuhn, 1976).

The overlapping of the molecules in the native fibre leaves gaps between the ends of the molecules, called 'hole' zones (*Figure 1.5*). It is in these areas that

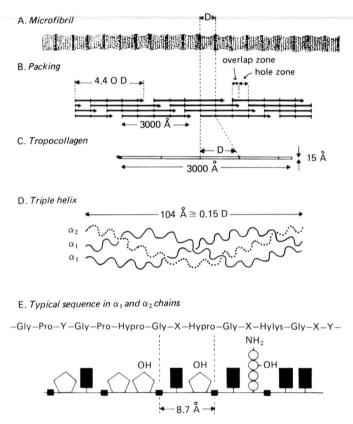

Figure 1.5. The structure of collagen. The diagram shows the relation-ship between the triple-stranded collagen molecules and the regular banding appearance which their overlapping arrangement produces. 10Å = 1 nm. (Reprinted by permission from Grant and Prockop (1972), New England Journal of Medicine, 286)

the first crystals of mineral may be laid down in the process of mineralization (*Figure 1.11*). The arrangement of collagen fibres may differ from one tissue to another; thus in normal bone matrix the fibres may be laid down as regular lamellae but in rapidly forming bone as woven fibres in all directions. Boyde (1972) provides a fascinating account of the 3-dimensional orientation of collagen fibres in bone, seen by scanning electron microscopy.

(b) Synthesis

Each polypeptide (α) chain is initially synthesized within the fibroblast or osteoblast as a precursor or pro α chain, with long extensions at both ends of the molecule which are subsequently removed.

This process begins with the translation of messenger RNA (mRNA) on membrane-bound ribosomes *(Figure 1.6)*. Following this certain post-translational modifications occur which include hydroxylation of susceptible prolyl and lysyl residues, and also glycosylation of some hydroxylysyl residues, utilizing specific enzymes. Molecular assembly then occurs, with alignment of the appropriate 3 polypeptide chains, aided by their extensions, followed by disulphide bound

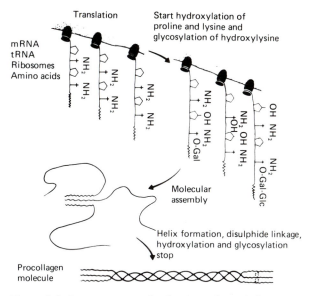

Figure 1.6. Important steps in the formation of the pro-collagen molecule (From Pinnell, 1978)

formation. Helix formation follows and it is this complete triple helical precursor molecule, procollagen, which is exported from the cell. To aid understanding of the effects of inherited disease it is usual to separate the post-translational modifications into separate steps, but this may be artificial. Thus hydroxylation and glycosylation may continue during molecular assembly, although they stop when the helix is formed. Some features of the pro α chain are shown in *Figure 1.7*; its exact structure is still being elucidated and this diagram merely outlines its different regions, distinguishing the large disulphide-containing extensions from the central helical region.

Once outside the cell the extensions of the procollagen molecule are removed by specific peptidases and fibril formation with cross-linking begins.

Mature collagen owes its physical properties to these cross-links, but only a minority of them have been identified. Important amongst the identified cross-links are those which involve hydroxylysine or lysine residues and their aldehyde derivatives. The aldehydes are produced by the action of a copper-dependent enzyme lysyl oxidase, and links may be formed between the aldehyde of lysine and the amino group of hydroxylysine, (referred to as a Schiff base) or two aldehyde groups (called an aldol condensation). The cross-linking of elastin also depends on the formation of these aldehydes from lysine and hydroxylysine.

13

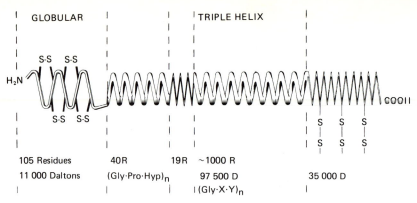

GLOBULAR   TRIPLE HELIX

S-S S-S

S-S S-S   S   S   S
                                            S   S   S

105 Residues   40R   19R   ~1000 R
11 000 Daltons   (Gly·Pro·Hyp)n   97 500 D   35 000 D
                                 (Gly·X·Y)n

Figure 1.7. A diagram of the structure of the pro α chain of collagen demonstrating extensions at both ends of the chain which are subsequently removed by peptidases

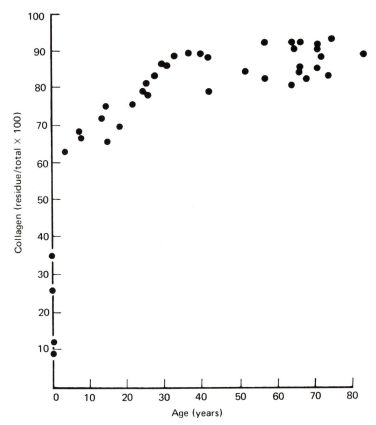

Figure 1.8. Changes in the stability of cross-linked polymeric collagen extracted from human skin. With increasing age the residual insoluble collagen after exposure to such agents as cold alkali increases, implying an increase in cross-linking. The points relating to subjects below one year of age should be particularly noted. (From Francis, Smith and MacMillan, 1973)

14

Cross-links of collagen may be intramolecular and intermolecular. Progressive intermolecular cross-linking occurs with age up to about 30 years (*Figure 1.8*), and such cross-linking also occurs with the maturation of young collagen in scar tissue.

There is also an alteration in the genetic type of collagen with age; for instance, the ratio of Type III to Type I collagen in the skin falls progressively until adult life (*Figure 1.9*) and scar tissues initially contain an excess of Type III collagen.

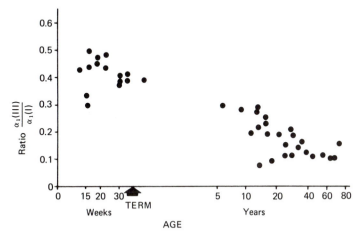

Figure 1.9. The reduction of Type III collagen relative to Type I in the skin with increasing age. The ratio is expressed in terms of the specific α chains derived from these collagens. (From Sykes *et al.,* 1976. Copyright by Academic Press Inc.)

Some of the disorders we shall subsequently discuss arise from genetic abnormalities in collagen synthesis (Chapters 7 and 8), and have been summarized by Bailey, Robins and Balian (1974), McKusick and Martin (1975), and Pinnell (1978). Knowledge of the structure and synthesis of collagen is constantly being revised.

(c) Breakdown

The mechanisms by which native collagen is degraded have been difficult to establish (Harris and Krane, 1974), but it seems that the initial breakdown is mostly brought about by collagenases similar to those first described in the tadpole. This enzyme splits the collagen molecule transversely into two fragments of ¼ and ¾; at temperatures above 32°C, including normal human temperature, the fragments thermally denature spontaneously and are further degraded, either by extracellular neutral proteases or by intracellular phagocytosis with lysosomal enzymes.

When collagen is broken down the hydroxyproline and hydroxylysine formed during collagen synthesis cannot be reutilized to make further protein and is either excreted in the urine, mainly as peptides, or oxidized. Although the percentage of hydroxyproline oxidized is considerable it appears to be fairly constant and therefore the hydroxyproline in the urine gives a good indication of that released from collagen.

In a patient on a low gelatin diet (gelatin is denatured collagen), the total amount of hydroxyproline (THP, free and peptide-bound,) in the urine of an adult is less than 50 mg a day, and increases when collagen turnover increases. Since bone contains so much of the collagen of the body and bone collagen is metabolically more active than most other tissue collagens, THP excretion provides a reliable index of bone turnover.

Most hydroxyproline in the urine is excreted as small peptides, but some peptides, which comprise 1 to 10 per cent of total hydroxyproline, may be very large, and these polypeptides have been related to the synthesis of new collagen rather than to its breakdown. Total hydroxyproline excretion is increased in a number of physiological and pathological states (Prockop and Kivirriko, 1967) such as in the rapid growth of adolescence (Clark, 1977), in Paget's disease (Chapter 5), and in hyperparathyroidism (*see also* Chapter 2).

Interestingly the distribution of hydroxylysine fractions in the urine appears to be different. Only 10 per cent of hydroxylysine is peptide-bound, none of this is dialysable, and analysis shows it to be highly glycosylated (Askenasi, Rao and Devos, 1976).

Krane *et al.* (1977) took advantage of the differing glycosylation of skin and bone collagen to infer the origin of glycosylated hydroxylysines in urine. In patients with active Paget's disease the ratio of glycosylgalactosyl hydroxylysine to galactosyl hydroxylysine in the urine was close to that of normal (and Pagetic) bone, and lower than that for skin. This ratio increased when bone resorption was decreased with calcitonin, and in normal subjects given calcitonin reached values much higher than for skin or bone. It was suggested that the additional source of collagen breakdown was the C1q subcomponent of complement.

Although it is usual to measure only urinary hydroxyproline there are also collagen-derived fragments in the blood. Thus the non-protein-bound hydroxyproline of plasma correlates well with urine hydroxyproline, but the protein-bound hydroxyproline does not.

(d) Examination of collagen

Changes in collagen may be inferred from alteration in the excretion or concentration of hydroxyproline peptides, but information may also be obtained from direct examination of tissues, and from connective tissue cells derived from them. Skin is often used as a source, since it is accessible and it is difficult to obtain undenatured collagen from bone. Measurements have been made on amino acid content, salt solubility, cross-linking (by direct and indirect methods) and more recently by acrylamide gel analysis of the genetic composition of collagen from pepsin-digested skin. The last method has been used to obtain the results for *Figure 1.9.*

The non-collagenous protein of bone matrix

Of the organic matter of adult bone 88 to 90 per cent is collagen and 10 to 12 per cent non-collagenous material, mainly protein. This non-collagen protein fraction contains a considerable number of components whose chemistry is still being worked out, and whose functions are largely unknown. It seems likely that it contains a group of substances of considerable physiological significance. The

16

reader should consult Herring (1968). Leaver, Triffitt and Holbrook (1975), Vaughan (1975) and Triffitt *et al.* (1976) to amplify the following account.

The non-collagenous proteins of bone may be divided into glycoproteins and proteoglycans (protein-polysaccharides), each of which are composed of a protein core or 'body' with carbohydrate 'arms', joined by linkage regions.

In general terms the arms of glycoproteins are short and may be branched, with varying sugars such as galactose, mannose, glucose, glucosamine and galactosamine, and sialic acid; whilst a proteoglycan (*Figure 1.10*) has long arms

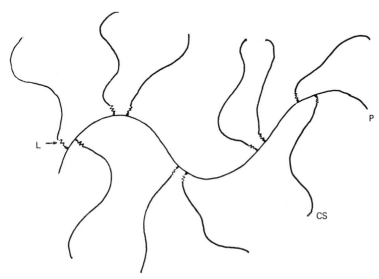

Figure 1.10. To show the general structure of a protein-polysaccharide (proteoglycan) molecule. P = *protein core;* CS = *chondroitin sulphate chain;* L = *linkage region.* (From Herring, 1968)

consisting of regularly repeating disaccharide units which may have sulphate groups attached to them and therefore form acid mucopolysaccharides (glycosaminoglycans). These glycosaminoglycans may be chondroitin-4-sulphate or 6-sulphate, or dermatan sulphate and keratan sulphate (*see also Figure 9.1*). Whilst work of most clinical relevance has been concerned with the enzymic defects responsible for the step-wise degradation of mucopolysaccharides (Chapter 9), studies have also been done on the glycoproteins of bone (such as sialoprotein), on glycoproteins which are synthesized elsewhere and found in bone (such as the α_2HS-glycoprotein), and on other plasma proteins in bone (such as albumin). Additionally a protein containing γ carboxyglutamic acid has been identified.

(a) Sialoprotein

This unusual glycoprotein isolated from bovine bone has a molecular weight of 23 000 with a single polypeptide chain with large amounts of aspartic and glutamic acid, and a large amount of sialic acid. It has a high metal-binding capacity.

(b) Glycoproteins of bone synthesised elsewhere

Chief amongst these is the α_2HS-glycoprotein which appears to be synthesized in the liver and to be concentrated in bone. Its concentration appears to be reduced where bone formation is increased, as in Paget's disease of bone (Ashton,

TABLE 1.II

The concentration of α_2HS-glycoprotein in plasma of patients with Paget's disease of bone compared with normal volunteers. (From Ashton, Hohling and Triffitt, 1976)

Subjects	Sex	α_2 HS-glycoprotein Conc. mg/100 ml	% normal
Normal	M	90 ± 6 (11)	100
	F	101 ± 11 (8)	100
Paget's	M	63 ± 10 (7)	70
disease	F	74 ± 15 (7)	74

Values mean ± SD. Number of individuals in brackets

Hohling and Triffitt, 1976, Table 1.II). Although bone resorption is also increased in this disorder, it is unlikely that α_2HS-glycoprotein will be released as such into the circulation.

(c) Albumin

Albumin has been shown to be a constituent of the organic matrix of bone. This is not in itself surprising since this protein presumably circulates through many tissues after leakage from the capillaries. However some of the albumin appears to be permanently incorporated into the calcified matrix (Owen and Triffitt, 1976).

(d) γ-Carboxyglutamic acid in protein

A protein which contains γ-carboxyglutamic acid (Gla) has been identified in bone (Price, Poser and Raman, 1976), and is of considerable interest. Carboxylation of the γ position of glutamic acid gives it considerable affinity for calcium, and this has been well demonstrated for the vitamin-K-dependent blood-clotting factors. The bone protein differs from these and contains hydroxyproline, suggesting that it is made by the same cells which synthesize collagen. A similar, if not identical, protein has been found in tissues which have undergone ectopic calcification, in scleroderma, dermatomyositis and atheromatous aorta (Lian et al., 1976). The Gla-containing protein from bone binds strongly to hydroxyapatite crystals but not to amorphous calcium phosphate, and strongly inhibits the formation of hydroxyapatite crystal nuclei. These effects are analogous to those of the phosphonates.

Bone mineral

The mineral of bone may be amorphous or crystalline. The amount of the amorphous phase of calcium phosphate is greatest in new bone and in bone from young animals. The crystals, when examined *in vitro,* are composed of hydroxyapatite $Ca_{10}(PO_4)_6(OH)_2$, but several other ions can substitute within the crystal lattice or be absorbed on its surface. These include fluoride, strontium, sodium, magnesium, copper, zinc, lead and radium. The substitution of many of these ions has clinical significance. Thus fluoride (in small amounts) may improve the stability of the crystal; radium and isotopes of strontium may produce a cumulative radioactive hazard; and lead may accumulate as a poison. The amorphous phase may be of considerable physiological importance; it is not of constant chemical composition and its equilibrium with the crystalline phase may be altered by many factors, including renal failure. Potts and Deftos (1974) state that the bones contain 99 per cent of the body's calcium, 85 per cent of the phosphorus, and 66 per cent of the magnesium.

Mineralization

The deposition of mineral on the matrix of bone is the one feature which distinguishes it from all other tissues, but the way in which this occurs is unknown. There is no shortage of theories, each with their own supporters. Again if we consult Vaughan (1975) and Rasmussen and Bordier (1974), the emphasis placed on the possible factors is very different: the first author emphasizes the importance of the protein-carbohydrate components of the bone matrix, at the expense of collagen—mineral interactions; the second authors consider that collagen—mineral interactions are of considerable importance, and also stress the cellular control of mineralization.

A number of features must be taken into account. These are: (1), the bone fluid; (2), the bone matrix; (3), the bone cells; and (4), possible inhibitors. The following brief account may be of use.

Bone fluid.   We must recall that the bone surfaces are exposed to a specialized compartment of the extracellular fluid (ECF), the bone fluid, which appears to differ from the general ECF in composition and, in particular, is said to have a higher content of potassium, and a lower content of calcium, magnesium and sodium (Vaughan, 1975). Thus, whether or not the general ECF is supersaturated with respect to bone crystal may be irrelevant. Whatever fluid is considered it is important to distinguish whether the concentrations of Ca and P are high enough to initiate mineralization or merely to maintain it. It should be emphasized that despite the widespread acceptance of the existence of bone fluid, it has not been directly obtained or analysed.

Bone matrix.   Much work has shown that native collagen provides an ideal matrix for mineralization and that this may start in the 'hole zones' of the Petruska and Hodge (1964) model of collagen. Calculations suggest that 50 per cent of total bone mineral may be accommodated in these 'holes', with a continuous mineral phase from one microfibrillar unit to the next. The one-dimensional diagram of collagen structure shown in *Figure 1.5* demonstrates how these hole zones occur; since the collagen fibril is a three-dimensional structure (*Figure 1.11*), the available space for mineral is considerable.

19

Collagen is not calcifiable immediately after its formation, but becomes so within a few days. Mineral is then rapidly deposited in an area recognizable histologically as the calcification front. It has been suggested that phosphate is important for this initial collagen—mineral interaction.

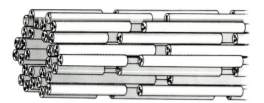 

Figure 1.11. A diagram to show how the three-dimensional structure of collagen can provide room for a continuous mineral phase based on the gaps or 'hole zones' between consecutive collagen molecules. Each cylinder represents a triple-stranded collagen molecule. The arrangement of the molecules within the micro-fibril is not accurately known. (Part of a figure from Rasmussen and Bordier, 1974)

The non-collagen proteins may also be important, but their mode of action is unclear. They may be effective in mineralization because of their metal-binding properties, particularly marked for sialoprotein. α_2HS-glycoprotein also appears to be related to mineralizing bone (Ashton, Hohling and Triffitt, 1976). The possible importance of a protein containing γ-carboxyglutamic acid has been referred to.

Bone cells.   The osteoblasts and the so-called calcifying vesicles demonstrate that cellular activity is essential for mineralization. Views differ on how much the osteoblast does or can do. It is suggested that in addition to making collagen early in its life, it also packages bone mineral (and possibly enzymes) which are exported into the newly formed osteoid to initiate mineralization.

The existence of calcifying vesicles has been most convincingly demonstrated histologically in epiphyseal plates (Ali, Sajdera and Anderson 1970; Ali, 1976), and vesicles have been isolated from collagenase-digested cartilage. These vesicles probably accumulate calcium and form hydroxyapatite crystals. They may be shown to contain such crystals as well as to have the important enzyme properties of an alkaline phosphatase, ATPase and pyrophosphatase (Anderson, 1976). Similar vesicles have been demonstrated in mineralizing bone (Gay and Schraer, 1975).

Inhibitors.   Since non-bone collagen does not mineralize but bone collagen does, and since calcification is not a universal feature of tissues despite the high concentrations of calcium and phosphorus in the extracellular fluid (sufficient to maintain if not initiate mineralization), it has been postulated that there are important mineralization inhibitors normally present. One of these may be pyrophosphate (Russell and Fleisch, 1976) which inhibits the precipitation and formation of calcium phosphate crystals *in vitro* and when given parenterally prevents induced ectopic mineralization in animals. Since pyrophosphate is normally present in the body fluids, there must be some mechanism for its destruction before calcification can occur. This is likely to be brought about by a pyrophosphatase (which is also an alkaline phosphatase) known to be associated with calcification. Where there is a deficiency of alkaline phosphatase (and hence pyrophosphatase) one might expect this to produce a disorder of mineralization. This is in fact seen in hypophosphatasia (Chapter 11) where there is an increase in plasma and urinary pyrophosphate and bone disorder superficially resembling rickets.

However the extent to which pyrophosphate is important in mineralization has yet to be established. For further details the reader should consult Russell (1975).

CONTROL OF BONE COMPOSITION

In the last 30 years (Albright and Reifenstein, 1948), the control of the composition of bone has become virtually synonymous with 'mineral homeostasis', since the mineral component of bone has been so much easier to examine than the matrix. This has led to considerable advances in our understanding of calcium metabolism in general and of the metabolism of vitamin D, parathyroid hormone and calcitonin in particular. At the same time it has effectively suppressed any wider consideration of the control of other components of bone.

Here we will consider those hormones which primarily control mineral metabolism and particularly calcium and phosphate. The reader will find that some of the points are repeated or amplified in the physiological sections of the appropriate chapters. This is intentional, because the points which are made are essential for the understanding of the clinical disorders. It is helpful first to have an outline of the main features of calcium (and phosphorus) balance in the adult.

Calcium and phosphorus balance

It is not possible to give a full description of calcium homeostasis because the facts are not known, and in several respects, particularly calcium exchange with the skeleton, the picture is far more confused than it was. Potts and Deftos (1974) give a detailed account of calcium metabolism, and Parfitt (1976) considers models of calcium homeostasis. *Figure 1.12* gives approximate figures for calcium exchanges between different organs, and a similar diagram is given by Paterson (1974); it also suggests that there is a considerable exchange between the general ECF and bone fluid. The reader should bear in mind Albright and Reifenstein's (1948) comment and be 'cautioned against taking these diagrams too literally'.

The classic features of calcium balance in the normal adult shown in *Figure 1.12* can be modified by conversion of the units to mmol (for calcium 1 mmol = 40 mg), and by recognition of the importance of calcium exchange between bone tissue fluid and extracellular fluid. The average adult takes in approximately 25 mmol of calcium per day in the food, of which the net absorption is 5 mmol, the remaining 20 mmol being lost in the faeces. The net absorption is the result of absorption minus secretion of calcium into the gut. The absorbed calcium contributes to the plasma calcium which circulates in a concentration of 9 to

10.4 mg per 100 ml (2.25 to 2.55 mmol per litre). This equilibrates with some of the calcium in the skeleton (which totals about 1kg; 25 000 mmol), and there is a continual process of incorporation of calcium into new bone mineral and liberation of calcium from resorbing bones. Very approximate figures for this are 10 mmol a day for each. Although the indirect results of kinetic studies

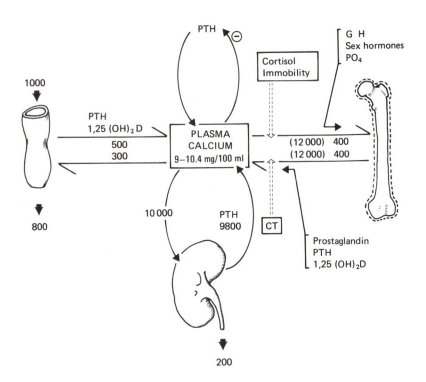

Figure 1.12. An outline of calcium homeostasis. Recent work has made a simple and correct diagram impossible. Arrows indicate the known routes of calcium exchange and the appropriate amount per day in mg (1 mmol = 40 mg for Ca). The effects of various hormones are indicated. The envelope surrounding the bone indicates the separation of bone fluid from the general ECF. There are uncertainties about the magnitude of exchange between bone fluid and ECF and relative importance of bone and kidney in normal homeostasis. Dotted arrows indicate blocks in formation or resorption of bone. (Numbers in brackets are approximate.) There is also an exchange of calcium resulting from bone formation and resorption of some 400 mg (= 10 mmol) per day. CT, Calcitonin; GH, Growth hormone. Thyroxine (not shown) increases collagen turnover, with bone resorption in excess of formation

suggest that only a small fraction of skeletal calcium is available for exchange, and the amounts involved in the normal processes of bone formation and resorption are very small in comparison to those handled by the kidney, there appears to be a very rapid and important exchange of calcium between bone fluid and ECF, independent of bone resorption and formation, which has been put as high as 300 mmol per day (Reeve, 1977).

The other important tissue in calcium homeostasis is the kidney. Each day about 250 mmol of calcium are filtered through the kidney and 245 mmol reabsorbed. For a patient in calcium balance, disregarding loss of calcium through sweat, the amount excreted in the urine, 5 mmol (200 mg) a day, equals that absorbed through the small intestine.

Clearly, in a growing person these figures are different and the net balance will be positive. As far as this simple model is concerned, the main arguments centre around the relative importance for calcium homeostasis of the processes occurring in the skeleton and in the kidney in health and disease. Whilst the daily control of plasma calcium is likely to be carried out by alterations in the large fluxes between the kidney, the extracellular fluid and the bone tissue fluid, the contribution of bone resorption and bone formation will be important where bone turnover is rapid.

Calcium in the plasma is either ionized (47 per cent of total calcium) or protein-bound (46 per cent), or complexed. The relationship between these depends on changes in pH and plasma-protein concentration. Ionized calcium is the component which is important in the control of PTH and calcitonin secretion.

Far less is known about the homeostasis of phosphate, and in the past its absorption across the intestine has been thought to follow passively that of calcium. The three factors which determine plasma phosphate are (Walton and Bijvoet, 1975): (1) The net inflow of phosphate into the extracellular space from gut, bone and soft tissues; (2) the tubular reabsorption of phosphate, which determines the average plasma phosphate at which phosphate inflow equals urinary excretion rate; and (3) glomerular filtration rate, which determines the change in plasma phosphate produced by a change in phosphate inflow. In steady state conditions with a normal GFR, the main determinant of plasma phosphorus is the rate of reabsorption by the kidney, which is particularly influenced by parathyroid hormone. There are inherited disorders both with low and high tubular reabsorption of phosphate (hypophosphataemic rickets and tumoral calcinosis, respectively). There is some evidence that calcitonin may be particularly important in the control of phosphate. It is important to remember that, phosphorus, like calcium, exists in the plasma in a number of forms and that more than half of it is organic. When we talk about phosphorus or phosphate in the plasma, we mean inorganic phosphate which is present in several diffusible forms some of which are protein-bound (Paterson, 1974). These and related points are dealt with in detail by Wilkinson (1976). It is interesting that phosphate itself appears to increase bone formation.

Hormones act at various points on mineral homeostasis. The major known effects are shown in *Figure 1.12* and dealt with in subsequent pages.

Vitamin D

The rapid advances in our knowledge of vitamin D metabolism have led to two important changes in our ideas about this substance. Firstly, that it requires conversion to active metabolites within the body before it can have its biological effect; and secondly, that of these metabolites, 1,25-dihydroxycholecalciferol ($1,25(OH)_2D$) can properly be considered as a hormone (rather than a vitamin) produced by the kidney according to the demands of its target organs and under close feedback control.

23

Vitamin D is normally made from precursors in the skin or taken by mouth. Naturally occurring vitamin D, cholecalciferol, made in the skin from 7-dehydrocholesterol, is called vitamin D_3; vitamin D_2, ergocalciferol, is made by irradiation of ergosterol, a plant sterol. According to its degree of fortification with vitamin D_2, food contains a varying proportion of D_2 and D_3. Oral vitamin

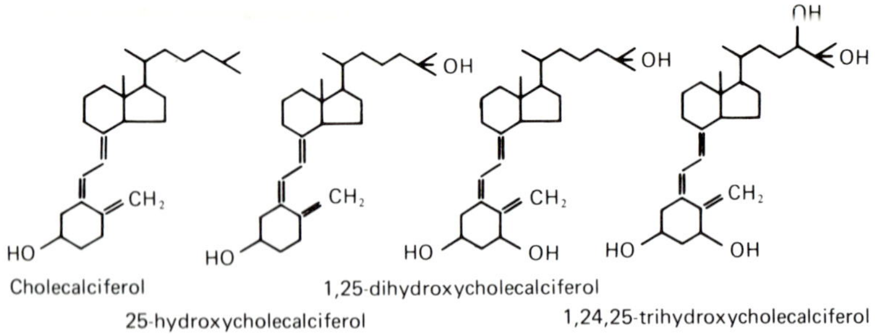

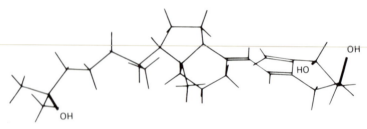

Cholecalciferol   1,25-dihydroxycholecalciferol

25-hydroxycholecalciferol   1,24,25-trihydroxycholecalciferol

Figure 1.13. The structure of vitamin D and some of its derivatives. (From Vaughan, 1975)

Figure 1.14. To demonstrate the three-dimensional structure of 1,25 dihydroxycholecalciferol. The 25-hydroxy position is to the left and the 1-hydroxy position on the right. (Diagram kindly supplied by Dr. David Fraser)

D is absorbed through the small intestine. The main dietary sources are fish, dairy products and margarine and the requirements per day are probably near to 400 iu. (10 μg) for children and 100 iu (2.5 μg) for adults. The structures of vitamin D and some of its important derivatives are shown in *Figure 1.13*.* Although these formulae are usually represented in one plane, it is important to remember that the molecule has a three-dimensional structure (*Figure 1.14*).

The further metabolism of vitamin D (either D_2 or D_3) has been well described by Kodicek (1974) and subsequently by many others (*see* Haussler and McCain, 1977). Key steps (*Figure 1.15*) are first a 25-hydroxylation by microsomal enzyme systems in the liver and then an important 1 α-hydroxylation (at the

---

*The nomenclature of vitamin D metabolites differs according to the author. The main alternatives are: for 25-hydroxycholecalciferol – 25 HCC or 25(OH)CC or 25(OH)D; and for 1,25-dihydroxycholecalciferol – 1,25 DHCC, 1,25 (OH)₂D or 1,25 (OH)₂CC. The term α, as in 1α25(OH)₂D, refers to the stereochemical position of the hydroxyl group at the 1 position of the A ring of the vitamin D molecule and if omitted, can be assumed. It is usually used in describing 1α-hydroxylated vitamin D (i.e. the synthetic derivative without 25-hydroxylation) as 1α HCC or 1α (OH)D.

24

opposite end of the molecule from the 25 position) which occurs exclusively in the mitochondria of kidney tissue. This produces the most active metabolite of vitamin D, 1,25 $(OH)_2D$, whose daily production rate (to judge by its effect in various forms of rickets) may be about 1 μg a day.

The control of 1 α-hydroxylation is important. Probably not all the factors have been identified but some of them are shown in Table 1.III. There is good evidence (Haussler and McCain, 1977) that 1 α-hydroxylation is stimulated by

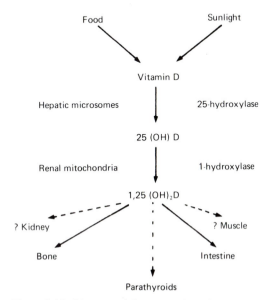

Figure 1.15. Diagram of the conversion of vitamin D to its metabolites. Vitamin D_3, cholecalciferol, is made in the skin from its precursor 7-dehydrocholesterol by the effect of ultraviolet light. Food also contains vitamin D_2, ergocalciferol, derived from irradiation of ergosterol. The subsequent metabolisms of D_2 and D_3 appear to be identical

increased concentrations of parathyroid hormone (related to low concentrations of calcium) and by hypophosphataemia. Animal experiments further indicate that increases in circulating calcium (reducing PTH) and phosphate will suppress 1 α-hydroxylation and allow the formation of the less active dihydroxylated derivative, 24,25$(OH)_2D$. These effects form the basis of a control system of considerable complexity.

In normal circumstances the circulating metabolites of vitamin D are 25(OH)D,24,25 $(OH)_2D$, and 1,25 $(OH)_2D$; there is also the trihydroxy derivative 1,24,25$(OH)_3D$. The first 3 of these are present in the plasma in approximate ratios of 1000:100:1.

Of the effects of vitamin D (via its active metabolites) on its target organs, the best known is the increased transport of calcium across the small intestine. In part this results from the stimulating effect of 1,25 $(OH)_2D$ on the synthesis of calcium binding protein by the intestinal cell (acting in this way like a steroid

TABLE 1.III

Factors thought to stimulate 1α-hydroxylation of 25(OH)D. (*See* Haussler and McCain, 1977)

	Factor	Comments
Likely	Low plasma phosphate	Direct action on kidney
	Low plasma calcium	By increasing PTH
	Increased PTH	? direct on kidney
Possible	Prolactin	Increased in lactation
	Oestrogen	Increased in pregnancy
	Cortisol	Opposite to effect on calcium transport
	Growth hormone	Effect obscure

hormone), but it appears that calcium transport may increase independently of such synthesis. There is also an increase in the activity of intestinal alkaline phosphatase and ATPase.

The only demonstrable effect of $1,25(OH)_2D$ on bone is an increase in resorption, and in fact this hormone is one of the most potent bone-resorbing agents known. Parathyroid hormone may owe some of its actions on the small intestine and bone to its stimulation of 1 α-hydroxylation. Certainly PTH has little effect on bone in the absence of vitamin D. This increase in resorption does not explain why the main clinical result of giving vitamin D (or $1,25(OH)_2D$) to patients with rickets or osteomalacia is to restore mineralization of bone, although the response to vitamin D itself could be due to the combined actions of a number of metabolites. It is possible that such mineralization is merely due to the increase in local concentrations of calcium and phosphate in the bone fluid, but more likely (although unproven) that it also results from the direct effect of vitamin D on the bone.

When vitamin D is given to the intact vitamin-D-deficient animal the renal reabsorption of phosphate is increased, but this is probably due to a reduction in PTH. However it is likely that $1,25(OH)_2D$ has a direct effect on the kidney since its production is controlled by the renal 1α-hydroxylase.

Amongst other possible target organs for vitamin D, muscle has been considered. There is some evidence for this (Chapter 4), but the results are far from clear. Young, Brenton and Edwards (1977) suggest that the weakness of vitamin D deficiency is due to muscle wasting.

It is clearly necessary to keep an open mind about other possible target organs of vitamin D, since there are many processes, such as endocrine secretion and motor-end plate function, which depend on appropriate concentrations of calcium, and many tissues contain calcium-binding proteins. Unexpectedly some data also suggest that vitamin D deficiency and its subsequent correction may alter the degree of cross-linking of collagen.

Despite the therapeutic developments which have followed the discovery of $1,25(OH)_2D$ (Chapter 4), we must not neglect the other metabolites. In particular there is evidence that $24,25(OH)_2D$, which normally circulates in a higher concentration than $1,25(OH)_2D$, increases calcium absorption when given in microgram doses even to nephrectomized man (Kanis *et al.*, 1978). Haussler and McCain (1977) provide an extensive and up-to-date review of vitamin D metabolism to which the reader is referred for further details.

Parathyroid hormone (PTH)

This polypeptide hormone, which consists of a chain of 84 amino acids, is derived from precursors which include proparathyroid hormone, and is secreted from the gland as a whole molecule but subsequently circulates as fragments, only some of which (the N-terminal fragments) are biologically active.

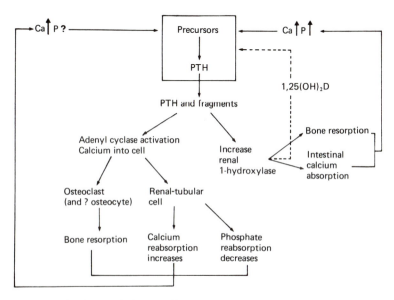

Figure 1.16. The synthesis and effects of parathyroid hormone. It is not clear which of the physiological effects of PTH require the activation of adenyl cyclase (see Parfitt, 1976, and Haussler and McCain, 1977). Note that all the effects of PTH tend to correct the hypocalcaemia which stimulates its production. The separate effects on phosphate are not all in the same direction, although the overall result of PTH excess is hypophosphataemia

The biological activity of parathyroid hormone depends on the N-terminal part of its molecule, and may be maintained after removal of residues from the carboxyl end. Tomlinson and O'Riordan (1978) point out that the initial product of biosynthesis has an additional 31 amino acids at the amino terminal end of the intact 1–84 peptide, which are subsequently removed in two stages. They also point out that the hormone fragments are not generated in the plasma, but result from the metabolism of PTH by kidney and liver.

PTH secretion increases when plasma-ionized calcium falls, and the effects of PTH are to increase the intestinal absorption and renal reabsorption of calcium and the resorption of bone (*Figure 1.16*). All these effects will increase plasma levels of calcium, which will in turn switch off PTH secretion. There is some evidence that parathyroid activity is also affected by 1,25 $(OH)_2D$.

PTH also has independent renal effects; it reduces reabsorption of phosphate, and thereby lowers plasma phosphate; it reduces the reabsorption of bicarbonate and amino acids, tending to produce a reversible hyperchloraemic acidosis and amino-aciduria.

The cellular actions of PTH on bone and kidney are mediated by an increase in the production of cyclic 3'5' adenosine monophosphate (cyclic AMP) but the details of its effect on the intestine are not clear. This cyclic AMP mechanism is of fundamental importance in the understanding of hormone action (Peacock, 1976). It is postulated that for PTH, and for many other peptide hormones, the hormone is bound at specific cell membrane receptors of the target organ and activates the enzyme adenyl cyclase. This produces cyclic AMP (from ATP) which becomes the 'second messenger' in the hormone's action, and may result in increased movement of calcium into the cell. The target cell for PTH in bone is the osteoclast and stimulation of its activity is the main way in which PTH increases bone resorption. Whether PTH also produces an increase in osteocyte activity with osteolysis is not agreed upon. The actions of parathyroid hormone on bone and its cells are discussed in detail in the four-part review by Parfitt (1976). In small amounts, and under certain conditions, both PTH and N-terminal fragments may stimulate bone formation. The administration of PTH normally causes a considerable increase in the plasma concentration and urinary excretion of cyclic AMP. The presence or absence of this increase after PTH can be a useful diagnostic test in hypoparathyroid states. The effects of metabolic blocks producing parathyroid insufficiency are further discussed in Chapter 6.

One should note that it is not at present clear how much the effects of PTH depend on its stimulation of the renal-1-hydroxylase.

Calcitonin

Calcitonin is a polypeptide of 32 amino acids with a disulphide bridge at one end; it is secreted by the C-cells of the thyroid which appear to have a common origin in neuroectoderm with other polypeptide-secreting cells. Calcitonins of porcine, salmon and human origin have many differences in amino acid content, with only nine residues common to all three. In contrast to PTH the secretion of calcitonin is stimulated by hypercalcaemia (*Figure 1.17*), and may be very high in patients with medullary carcinoma (C-cell carcinoma). In man its normal physiological function is unknown (Queener and Bell, 1975), and there is controversy about the methods used for its measurement. Recent work suggests that it may have a role in the bone disease of renal failure (Kanis *et al.* 1977).

Calcitonin acts on bone and kidney via adenyl cyclase receptors and reduces osteoclastic bone resorption *in vivo* and in culture. Its main clinical effect is seen where bone is turning over rapidly, as in growth or in Paget's disease. Calcitonin has additional effects on the kidney causing a natriuresis and increased excretion of phosphate. The secretion of calcitonin can be stimulated by alcohol and by glucagon and gastrin (thus levels of both gastrin and calcitonin are high in the Zollinger–Ellison syndrome). Physiologically the main stimuli to calcitonin secretion may come from the gastrointestinal tract.

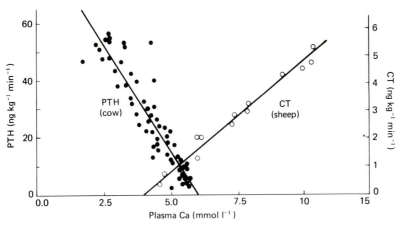

Figure 1.17. To show the contrasting effects of alteration of plasma calcium on the secretion rate of PTH and calcitonin (From Copp, 1969)

Other hormones

Many other hormones affect mineral homeostasis but their major action is on the structure of the skeleton, and they are thus concerned with skeletal homeostasis. There is controversy about their actions, which the following description takes into account. Jowsey (1977) reviews their morphological effects.

Growth hormone

It is now considered that most of the effects of growth hormone are mediated by a group of growth-hormone-dependent peptides called somatomedins, which appear to be manufactured by the liver. Thus growth hormone itself has no *in vitro* effect on cartilage chondrocytes whereas the somatomedins have. *In vivo,* growth hormone presumably owes its effects on the skeleton to the production of somatomedins. Normally the secretion of growth hormone is intermittent throughout the 24 hours, and is stimulated by factors such as insulin, apprehension and sleep. It is not increased during the rapid physiological growth of adolescence. Growth hormone secretion is controlled by hypothalamic-releasing and-inhibiting factors and somatomedins probably also exercise some control over its production. Thus in children who are not able to produce somatomedins (owing to an inherited biochemical defect) the growth hormone level is high. A recent study (Schwalbe *et al.,* 1977) showed that compared with normal the somatomedin activities in children with growth hormone deficiency were low and in those with sexual precocity were high. The somatomedin activity was low in chronic renal failure. Van den Brande (1976) describes situations in which linear growth and somatomedin levels are not correlated. Thus somatomedin levels in the plasma are low during the rapid growth of recovery from the dwarfism of emotional deprivation (p. 67) and are normal in achondroplasia. The sensitivity of tissues to somatomedin probably alters with age and in disease. Its assay may be complicated by the presence of inhibitors and by binding to specific proteins.

It is not exactly clear how somatomedins produce their effects, especially in the adult skeleton. They stimulate the uptake of ^{35}S into the cartilage of hypophysectomized rats, and have a similar effect on cultured chondrocytes. Mitogenesis is stimulated not only in the epiphyseal cartilage cells but also in the cells of a number of soft tissues, such as the intestine, liver and pancreas. Although the main effects of somatomedins are on cartilage, they can also stimulate collagen synthesis, have biochemical effects on muscle and adipose tissue, and have insulin-like activity.

Clinically, excess of growth hormone (and hence somatomedins) before growth has finished produces gigantism, and after the epiphyses have fused, acromegaly. Likewise, growth hormone deficiency produces dwarfism. In acromegaly, the overall size of the skeleton increases, but at the expense of endosteal and trabecular loss of bone. There is new periosteal formation of bone and renewed formation of bone by the vertebral end-plates. Radiologically the bones may appear osteoporotic and bone mass may be normal or low. In addition to the changes in bone (Jowsey, 1977) there is an increase in articular cartilage, an early degenerative arthritis, and enlargement of the soft tissues and extremities. Biochemically the plasma phosphate may be high, owing to an increase in TmP/GFR (Chapter 2), and there may be hypercalcaemia (p. 165) and hypercalcuria.

It is not known whether all the effects of growth hormone can be attributed to the somatomedins which they generate, but this seems likely. However, it is entirely possible that once the biochemistry is sorted out, we shall be confronted with a situation like that of vitamin D metabolism, but with hormones which control structural rather than mineral homeostasis.

Thyroxine

Thyroxine and related hormones have effects on many tissues and that on the skeleton is often ignored. However there seems no doubt that in thyrotoxicosis there is an increased turnover of bone with resorption in excess of formation. This may cause a significant increase in urinary hydroxyproline, with hypercalcaemia, hyperphosphataemia and hypercalcuria. Where hyperparathyroidism coexists this will be suggested by a low plasma phosphate.

A deficiency of thyroid hormone has its most striking effect on the skeleton when it occurs in childhood (Chapter 2). Apart from the clinical appearance of hypothyroidism, which in any case may not be marked, the main features are growth retardation, and stippling of the epiphyseal centres with a delay in their appearance. In adults hypothyroidism has little clinical effect on the skeleton, but bone turnover is considerably decreased and hydroxyproline excretion reduced.

Sex hormones

Of the sex hormones, oestrogen is clinically the most important, because its deficiency probably contributes to postmenopausal osteoporosis (Chapter 3). Although a number of effects of oestrogen may be demonstrated in growing animals (Vaughan, 1975) little has been shown in man. Most results are compatible with the idea that postmenopausal oestrogen deficiency increases the sensitivity of bone to endogenous PTH, thus leading to increased bone resorption.

Deficiency of testosterone, either due to testicular failure or pituitary failure leads to osteoporosis.

Adrenal cortical hormones

An excess of adrenal corticosteroids, either occurring naturally as in Cushing's syndrome or more commonly given as treatment, may cause marked effects on the skeleton, with stunting of growth and loss of bone. There is an interference with the normal metabolic changes occurring in the cartilage matrix, a suppression of collagen synthesis and probably increased resorption of bone. In the adult skeleton an excess of adrenal corticosteroids is one of the known causes of osteoporosis.

Adrenal androgens are probably responsible for the spurt of adolescent growth. Early closure of epiphyses may be brought about by the administration of sex hormones, or their pathological excess as in the adrenogenital syndrome.

Local factors

Classical investigation of metabolic bone disease concentrates on the circulating hormones PTH, vitamin D and calcitonin, but there are other important substances with powerful local actions on bone. These include prostaglandins, and an osteoclast activating factor (OAF) which may have particular relevance in the bone resorption of myeloma.

Prostaglandins

A vast number of tissues produce prostaglandins (so-called because early research centred around prostatic tissues), which are a family of 20 carbon fatty acids with many effects (Samuelsson *et al.*, 1975). The series which have particularly attracted attention (prostaglandin E) are potent stimulators of bone resorption *in vitro* (Tashjian, 1975) and are thought to have a role in the effects of neoplasia on bone. Prostaglandins and the newly-described thromboxanes and prostacyclins are derived from short-lived intermediates, the cyclic endoperoxides. Although bone-resorbing activities are generally attributed to the prostaglandins, some of their biological effects could be due to these related substances or their metabolites. It is not yet clear how prostaglandins produce their effect on cells, although the cAMP system is certainly involved. Since prostaglandins are short-lived their effects are mainly local, and in this sense they cannot be classified as hormones. It is possible that they or their metabolites may also have some effect at a distance.

Certain animal tumours produce PGE_2 and resorb bone and as in man these activities may be reduced by indomethacin. The hypercalcaemia produced by bone metastases, as from breast cancer, is partly attributable to the effects of locally produced prostaglandins on bone. Such local effects may also contribute to hypercalcaemia even when metastases are not detectable (*Lancet*, 1976). It is

31

not always clear where the prostaglandins come from and experimental implantation of tumour cells into bone in animals suggests that bone resorption results from both the direct action of the tumour and from the stimulation of osteoclasts.

Osteoclast activating factor

When cultured normal human leucocytes are stimulated by an antigen or mitogen, they release an osteoclast activating factor (OAF). Lymphoid cell lines from patients with myeloma (also Burkitt's lymphoma and giant cell lymphoma) secrete a similar substance which may account for the increased resorption of bone (Mundy *et al.*, 1974).

SUMMARY

The main message which should emerge from this chapter is that the skeleton is not inert despite its appearance, and that it should be regarded as a metabolically active tissue whose composition is maintained by cells functioning under hormonal control within a specialized compartment of the extracellular fluid. Clearly much controversy exists about the function of bone cells, knowledge of bone mineral still outweighs that of matrix, and practically nothing is known about the control of bone structure. Despite these areas of ignorance sufficient is now known of bone physiology to help with the clinical disorders to be described.

REFERENCES

Aaron, J. E. (1976). Histology and microanatomy of bone. In *Calcium, Phosphate and Magnesium Metabolism*, pp. 298–356. Ed. B. E. C. Nordin. London: Churchill Livingstone

Albright, F. and Reinfenstein, E. C. (1948). *The Parathyroid Glands and Metabolic Bone Disease*. Baltimore: Williams and Wilkins Co.

Ali, S. Y. (1976). Analysis of matrix vesicles and their role in the calcification of epiphyseal cartilage. *Fedn. Proc. Fedn. Am. Socs exp. Biol.* **35**, 135–142

Ali, S. Y., Sajdera, S. W. and Anderson, H. C. (1970). Isolation and characterisation of calcifying matrix vesicles from epiphyseal cartilage. *Proc. natn. Acad. Sci. U.S.A.* **67**, 1513–1520

Anderson, H. C. (1976). Matrix vesicles of cartilage and bone. In *The Biochemistry and Physiology of Bone. Vol. IV: Calcification and Physiology*. Ed. Bourne, G. H. 2nd edition, pp. 135–157. New York and London: Academic Press

Ashton, B. A., Höhling, H. J. and Triffitt, J. T. (1976). Plasma proteins present in human cortical bone: Enrichment of the α_2HS-glycoprotein. *Calcif. Tissue Res.* **22**, 27–33

Askenasi, R., Rao, V. H. and Devos, A. (1976). Peptide-bound hydroxylysine and large polypeptides related to collagen synthesis. *Eur. J. clin Invest.* **6**, 361–363

Avioli, L. V. and Krane, S. M. (1977). *Metabolic Bone Disease*. Vol. 1. London: Academic Press

Bailey, A. J., Robins, S. P. and Balian, G. (1974). Biological significance of the intermolecular cross links of collagen. *Nature, Lond.* **251**, 105–109

Bassett, C. A. L. (1971). Biophysical principles affecting bone structure. In *The Biochemistry and Physiology of Bone.* 2nd Edition. Ed. Bourne G. H. Vol. III, pp. 1–76. London: Academic Press

Bornstein, P. (1974). The biosynthesis of collagen. *A. Rev. Biochem.* **43**, 567–603

REFERENCES

Bourne, G. H. (1972). *The Biochemistry and Physiology of Bone. Vol. III: Development and Growth.* 2nd edition. New York and London: Academic Press

Boyde, A. (1972). Scanning electron microscopic studies of bone. In *The Biochemistry and Physiology of Bone. Vol. I: Structure.* Ed. Bourne, G. H. 2nd edition. New York and London: Academic Press

Brookes, M. (1971). *The Blood Supply of Bone.* London: Butterworths

Clark, S. (1977). Longitudinal growth studies in normal and scoliotic children. In *Scoliosis.* Proceedings of a 5th Symposium, London, 1976. Ed. P.A. Zorab, pp. 165–180. London: Academic Press

Copp, D. H. (1969). Endocrine control of calcium homeostasis. *J. Endocr.* **43**, 137–161

Fietzek, P. P. and Kuhn, K. (1976). The primary structure of collagen. *Int. Rev. Connect. Tissue Res.* **7**, 1–60

Francis, M. J. O., Smith, R. and MacMillan, D. C. (1973). Polymeric collagen of skin in normal subjects and in patients with inherited connective tissue disorders. *Clin. Sci.* **44**, 429–438

Frost, H. M. (1964). *Mathematic Elements of Lamellar Bone Remodelling.* Springfield: Charles C. Thomas

Gay, C. and Schraer, H. (1975). Frozen thin sections of rapidly forming bone: bone cell ultrastructure. *Calcif. Tissue Res.* **19**, 39–49

Grant, M. E. and Prockop, D. J. (1972). The biosynthesis of collagen. *New Engl. J. Med.* **286**, 194–199, 242–249, 291–300

Hall, D. A. and Jackson, D. S. (Eds) (1976). *Int. Rev. Connect. Tissue Res.* Vol. 7. London: Academic Press

Harris, E. D., and Krane, S. M. (1974). Collagenases. *New Engl. J. Med.* **291**, 557–563, 605–609, 652–661

Haussler, M. R. and McCain, T. A. (1977). Vitamin D metabolism and action. *New Engl. J. Med.* **297**, 974–983, 1041–1050

Herring, G. M. (1968). The chemical structure of tendon, cartilage, dentin and bone matrix. *Clin. Orthop.* **60**, 261–299

Jowsey, J. (1977). *Metabolic Diseases of Bone.* Vol. 1. London: W. B. Saunders

Kanis, J. A., Smith, R., Cundy, T., Heynen, G., Bartlett, M., Warner, G. T. and Russell, R. G. G. (1978). Is 24, 25-dihydroxycholecalciferol a calcium-regulating hormone in man? *Br. med. J.* **1**, 1382–1386

Kanis, J. A., Earnshaw, M., Heynen, G., Ledingham, J. G. G., Oliver, D. O., Russell, R. G. G. Woods, C. G., Franchimont, P. and Gaspar, S. (1976). Bone turnover rates after bilateral nephrectomy: role of calcitonin. *New Engl. J. Med.* **296**, 1073–1079

Kodicek, E. (1974). The story of vitamin D. From Vitamin to Hormone, *Lancet,* **1**, 325–329

Lancet (1976). Osteolytic metastases. (Editorial). **2**, 1063–1064

Leaver, A. G., Triffitt, J. T. and Holbrook, I. B. (1975). Newer knowledge of non-collagenous protein in dentin and cortical bone matrix. *Clin. Orthop.* **110**, 269–292

Lian, J. B., Skinner, M., Glimcher, M. J. and Gallop, P. (1976). The presence of γ-carboxyglutamic acid in the proteins associated with ectopic calcification. *Biochem. biophys. Res. Commun.* **73**, 349–355

McKusick, V. A. and Martin, G. R. (1975). Molecular defects in collagen. *Ann. intern. Med.* **82**, 585–586

Mundy, G. R., Luben, R. A., Riasz, L. G., Oppenheim, J. J., and Buell, D. N. (1974). Boneresorbing activity in supernatants from lymphoid cell lines. *New Engl. J. Med.* **290**, 867–871

Neuman, W. F. (1969). The milieu interieur of bone: Claude Bernard revisited. *Fedn Proc. Fedn Am. Socs exp. Biol.* **28**, 1846–1850

Nimi, M. E. (1974). Collagen: its structure and function in normal and pathological connective tissues. *Semin Arthritis Rheum.* **4**, 95–150

Nordin, B. E. C. (1976). *Calcium, Phosphate and Magnesium Metabolism.* London: Churchill Livingstone

Owen, M. (1977). Precursors of osteogenic cells. 13th European Symposium on Calcified Tissues. *Calcif. Tissue Res.* (suppl) **24**, R 19

Owen, M. and Triffitt, J. T. (1976). Extravascular albumin in bone tissue. *J. Physiol.* **257**, 293–307

Parfitt, A. M. (1976). The actions of parathyroid hormone on bone: relation to bone remodelling and turnover, calcium homeostasis and metabolic bone disease. *Metabolism* **25**, 809–844, 909–955, 1033–1069, 1157–1188

33

REFERENCES

Paterson, C. R. (1974). *Metabolic Disorders of Bone*. Oxford: Blackwell Scientific

Peacock, M. (1976). Parathyroid hormone and calcitonin. In *Calcium, Phosphate and Magnesium Metabolism*. Ed. B. E. C. Nordin, pp. 405–443. London: Churchill Livingstone

Petruska, J. A. and Hodge, A. J. (1964). A subunit model for the tropocollagen macromolecule. *Proc. natn. Acad. Sci. U.S.A.* **51**, 871–876

Pinnell, S. R. (1978). Disorders of collagen. In *Metabolic Basis of Inherited Disease*. Eds. J. B. Stanbury, J. B. Wyngaarden and D. S. Fredrickson. 4th edn. pp. 1366–1394. New York: McGraw-Hill

Potts, J. T. and Deftos, L. J. (1974). Parathyroid hormone. Calcitonin, Vitamin D. Bone and bone mineral metabolism. In *Duncan's Diseases of Metabolism*. Eds P. K. Bondy, L. E. Rosenberg. London: W. B. Saunders. pp. 1225–1430

Price, P. A., Poser, J. W. and Raman, N. (1976). Primary structure of the γ-carboxyglutamic acid-containing protein from bovine bone. *Proc. natn. Acad. Sci. U.S.A.* **73**, 3374–3375

Prockop, D. J. and Kivirikko, K. I. (1967). Relationship of hydroxyproline excretion in the urine to collagen metabolism. Biochemistry and clinical applications. *Ann. intern. Med.* **66**, 1243–1267

Queener, S. F. and Bell, N. H. (1975). Calcitonin: a general survey. *Metabolism* **24**, 555–567

Rasmussen, H. and Bordier, Ph. (1973). The cellular basis of metabolic bone disease. *New Engl. J. Med.* **289**, 25–32

Rasmussen, H. and Bordier, Ph. (1974). *The Physiological and Cellular Basis of Metabolic Bone Disease*. Baltimore: Williams and Wilkins Co.

Reeve, J. (1977). Disorders of plasma calcium. *Hosp. Update.* **3**, 19–30

Russell, R. G. G. (1975). Diphosphonates and polyphosphates in medicine. *Br. J. Hosp. Med.* **14**, 297–314

Russell, R. G. G. and Fleisch, H. (1976). Pyrophosphate and diphosphonates. In *The Biochemistry and Physiology of Bone. Vol IV. Calcification and Physiology*. Ed. G. H. Bourne. 2nd edition. pp. 61–104. New York and London: Academic Press

Samuelsson, B., Granström, E., Green, K., Hamberg, M. and Hammarström, S. (1975). Prostaglandins. *Ann. Rev. Biochem.* **44**, 669–695

Schwalbe, S. L., Betts, P. R., Rayner, P. H. W. and Rudd, B. T. (1977). Somatomedin in growth disorders and chronic renal insufficiency in children. *Br. med. J.* **1**, 679–682

Sykes, B., Puddle, B., Francis, M. J. O. and Smith, R. (1976). The estimation of two collagens from human dermis by interrupted gel electrophoresis. *Biochem. biophys. Res. Commun.* **72**, 1472–1480

Tashjian, A. (1975). Prostaglandins, hypercalcaemia and cancer. *New Engl. J. Med.* **293**, 1317–1318

Tomlinson, S. and O'Riordan, J. L. H. (1978). The Parathyroids. *Br. J. Hosp. Med.* **19**, 40–53

Triffitt, J. T., Gebauer, U., Ashton, B. A., Owen, M. E. and Reynolds, J. J. (1976). Origin of plasma α2HS-glycoprotein and its accumulation in bone. *Nature Lond.* **262**, 226–227

Van den Brande, J. L. (1976). Plasma somatomedins. Clinical observations. In *Growth Hormone and Related Peptides*. Eds A. Pecile and E. E. Muller. Proceedings of the 3rd International Symposium, Milan, 1975. *Excerpta med.* pp. 271–285

Vaughan, J. (1975). *The Physiology of Bone*. 2nd edition. Oxford: Clarendon Press

Walton, R. J. and Bijvoet, O. L. M. (1975). Nomogram for derivation of renal phosphate concentration. *Lancet* **2**, 309–310

Wilkinson, R. (1976). Absorption of calcium, phosphorus and magnesium. In *Calcium, Phosphate and Magnesium Metabolism*. Ed. B. E. C. Nordin. pp. 36–112. London: Churchill Livingstone

Wootton, R., Reeve, J. and Veall, N. (1976). The clinical measurement of skeletal blood flow. *Clin. Sci. Molec. Med.* **50**, 261–268

Young, A., Brenton, D. P. and Edwards, R. H. T. (1978). Analysis of muscle weakness in osteomalacia. (Abstract). *Clin. Sci. Molec. Med.* **54**, 31

2

Diagnosis of Metabolic Bone Disease

INTRODUCTION

The sophisticated studies used in metabolic bone diseases have tended to obscure the usefulness of the history and physical examination, which still have an important part to play in diagnosis. Increasing neglect of clinical methods may also lead to unnecessary investigation, and incidentally makes the results of much biochemical research difficult to interpret. In this chapter the symptoms, signs and investigation of metabolic bone disease* will be described and compared. It is hoped that this will enable the reader who has a diagnostic problem to alight on the right chapter; the accompanying Tables (2.1–2.II) with the appropriate reservations in the text and in subsequent chapters, should help in this. The reader should particularly consult Fourman and Royer (1968), Paterson (1974), and Nordin (1976), for classic metabolic bone diseases and McKusick (1972) for inherited connective tissue disorders. In case he should disregard the importance of growth, this is dealt with first.

GROWTH

Growth may be abnormal in many diseases both of the skeleton and of other systems. The main features are short stature and disproportion, but excessive height may also cause problems (Tables 2.III–2.VI). The control of growth (skeletal homeostasis) is not well understood, but some hormonal influences such as the somatomedins have been dealt with in Chapter 1; and important changes in biochemical measurements during growth are shown in *Figures 2.1, 2.2* and *2.9*. A knowledge of the normal growth of the skeleton and the relationships between its different parts is essential for diagnosis (and treatment) in both

*The term metabolic bone disease is used here as a convenient shorthand (*see* Preface).

children and adults. The reader should refer to the classic work of Tanner (1973), the useful description of Vaughan (1975) and clinical descriptions of abnormal growth by Raiti (1969), Lacey and Parkin (1974) and Parkin (1975).

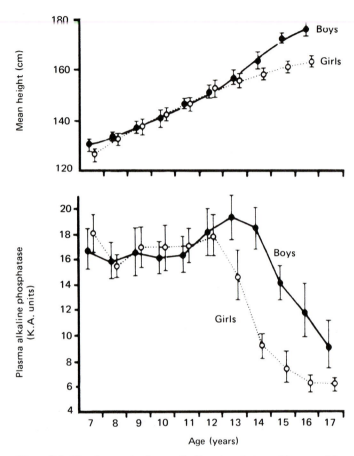

Figure 2.1. The changes in plasma alkaline phosphatase with age and its relation to height and sex. To convert to SI units (iu./l) multiply by 7.1. Bars indicate mean ± 2 S.E. of mean. (From Round, 1973)

The measurement of growth (particularly height) in children and its comparison with established normal ranges (expressed as standard deviations from the mean or as percentiles) is essential. Such comments as 'approximately normal height' or 'seems short' are not helpful. In addition to single measurements, sequential measurements with time enable growth velocity to be estimated, providing a valuable indication of spontaneous changes in growth and those changes induced by treatment. The most simple growth or percentile charts are those which record the normal range of height (or weight) throughout childhood, or those which show the normal rate of growth such as a velocity chart (*Figure 2.3*).

These are sufficient for most purposes,* but charts are also available which take into account the differing ages at which individuals mature and the different proportions (of sitting to standing height for instance) at different ages.

Of the normal changes in body proportion with increasing age the most important is that between the trunk and the limbs. Taking the division of the

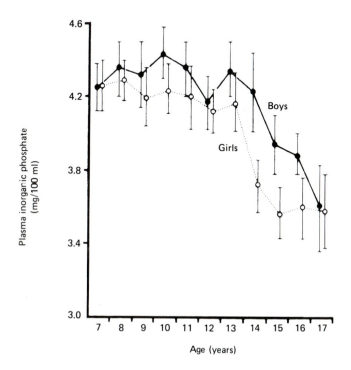

Figure 2.2. The changes in plasma inorganic phosphate with age. To convert to SI units (mmol/l) multiply by 0.32. (From Round, 1973)

upper and lower segments as the symphysis pubis, the ratio between them decreases from 1.7 at birth towards unity in late childhood (see Vaughan, 1975, Figure 10.5). These ratios may be abnormal for age in many conditions, particularly in the group of short-limbed dwarfism (such as achondroplasia and severe osteogenesis imperfecta). In the adult, knowledge of the normal relationship between the limbs and trunk may enable one to calculate loss of height (see also p. 47). The causes of short stature are discussed on page 66, and Tables 2.III–2.VI.

---

*Appropriate charts from Tanner and Whitehouse standards are available from Creaseys Ltd., Castle Mead, Hertford, U.K.

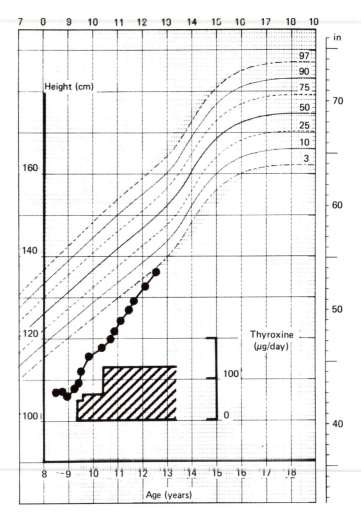

Figure 2.3. (a) The effect of thyroxine on the linear growth in a hypothyroid boy

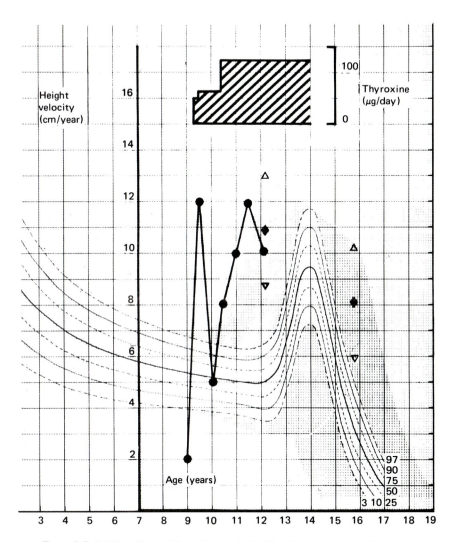

Figure 2.3. (b) The effect of thyroxine on the height velocity in a hypothyroid boy

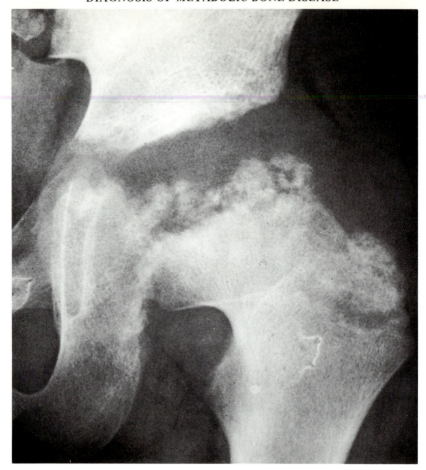

Figure 2.3 (c) X-ray of hips of the hypothyroid boy shows the characteristic stippled upper femoral epiphysis

CLINICAL FEATURES

History and symptoms

Pain, deformity and fracture are the common features of metabolic bone disease. To this may be added proximal myopathy (in osteomalacia and rickets) and the symptoms of any underlying disorder. Spontaneous tetany is rare (p. 48).

Pain

The cause of bone pain in metabolic bone disease is not well understood. Nerve fibres in bone are found mainly in the periosteum and around the blood vessels. Thus pain may follow stretching of the periosteum due to increased vascularity of bone, or to distortion and bending of the bone or to fracture.

The bone pain of osteomalacia is often generalized and accompanied by tenderness. The ribs particularly may be painful and tender to pressure. The generalized tenderness can be sufficient to keep a patient awake at night, and to cause him to avoid contact with other people. Looser's zones (pseudofractures, p. 107) in the long bones may cause additional local pain. In osteoporosis, tenderness of the spine is usually limited to the fractured vertebrae; where generalized bone pain and tenderness is encountered in a patient with vertebral collapse, some disease other than osteoporosis, such as multiple myeloma should be strongly suspected (*see* Chapter 3). It is rare for primary hyperparathyroidism to produce severe bone disease; when it does, pain is usually a feature which is rapidly relieved by parathyroidectomy. Pagetic bone may be painful, but the pain is often due to associated joint disease. In the absence of microfractures, persistent localized pain in a previously pain-free subject with known Paget's disease strongly suggests sarcoma. Except at the site of recent fracture, pain is not a feature of osteogenesis imperfecta. The characteristic bone pain of Engelmann's disease (Chapter 11) is inexplicable.

Deformity

The main deformity complained of in osteoporosis and osteomalacia is loss of height. Clothes no longer fit, since trunk height is reduced relative to the legs. Women particularly note that the waist of previously suitable clothes appears to have fallen. In osteomalacia the loss of height may be accentuated by proximal myopathy. One young subject had to give up driving because trunk weakness and loss of height made it impossible for him to see through the windscreen of his car. The deformities of active rickets in childhood include knock knees and bow legs, enlarged epiphyses and prominent bossing of the skull. In rickets the most severe dwarfism with misshapen long bones is seen in the vitamin-D-dependent form. Hypophosphataemic vitamin-D-resistant rickets is a cause of short-limbed dwarfism.

In Paget's disease the commonest deformity complained of is a thickened and bowed tibia. The generalized deformities illustrated in the earlier descriptions of this disease include a large tam-o'-shanter skull and bowing of the long bones, Those who still wear hats find it difficult to get a size big enough, and previously suitable spectacles no longer fit. Most patients with mild osteogenesis imperfecta do not develop any long-bone deformity, but have shortness of stature. In severe osteogenesis imperfecta ('OI congenita', Chapter 7) dwarfism may be extreme, the limbs are short in relation to the trunk, with bowing of the bones, and the head is large, particularly the vault, which is often misshapen. Bizarre alterations of growth, with asymmetry of bones and local hypertrophy should suggest osseous neurofibromatosis and, possibly fibrous dysplasia.

Fracture

The symptoms of fracture will clearly depend on the bone involved. In osteoporosis the commonest type of fracture is vertebral compression, with periodic localized severe pain and loss of height associated with each episode of compression. Fractures may also occur through Looser's zones, and by the extension of microfractures in Paget's disease. In severe osteogenesis imperfecta fractures may occur very easily with only slight pain.

Proximal muscle weakness

This is characteristic of osteomalacia (Chapter 4). Its cause is unknown. The first complaints are generally concerned with the lower limbs – a waddling gait, difficulty in climbing stairs, difficulty in getting out of a low chair. The legs may be described as stiff rather than weak. In rickets the child may go off its feet. Trunk weakness also occurs and there is weakness of the shoulder muscles, which may not be complained of unless the patient has a job which involves their use. However there may be difficulty raising the arms above the head and lifting objects off high shelves. These symptoms may also occur in various forms of myositis and in polymyalgia rheumatica (p. 115).

Underlying disorder

The symptoms of the underlying disorder are important, since they may provide important clues; they may for instance include nocturia (due to renal failure) steatorrhoea, symptoms of anaemia, of hypogonadism or of Cushing's syndrome.

Previous illnesses

It is of considerable importance to enquire about previous illnesses and operations. Information about a previous gastrectomy is unlikely to be witheld, but a hysterectomy is often considered as a medical episode not worth mentioning.

The family history

This is most important in inherited bone disorders. Important examples include osteogenesis imperfecta, the mucopolysaccharidoses and the inherited forms of rickets. One should remember that in some of these (for instance, Lowe's syndrome, hypophosphataemic rickets, Hunter's syndrome) inheritance may be X-linked. In others it may be dominant with variable expression (Marfan's syndrome, myositis ossificans progressiva, neurofibromatosis). Many systems may be involved in a dominantly inherited disease as in the multiple-endocrine adenoma syndrome (Chapter 6). Where siblings only are affected and the previous generations are apparently normal, this implies recessive inheritance, (for example Wilson's disease, hypophosphatasia). Although the inheritance of bone disorders often conforms to normal genetic patterns, one should be aware that the situation is not always simple, owing to variable suppression of one of the two X chromosomes in the female, to the existence of two different mutant forms of the same enzyme (as in some mucopolysaccharidoses) and to variable expression of apparently dominantly-inherited disorders. The rule that enzyme deficiencies are inherited as recessives and changes in structural proteins as dominants can only be accepted in the most general terms.

Other historical features

These will be suggested in subsequent chapters. Clearly it is necessary to know about such aspects as previous growth, change in appearance (acromegaly, Paget's disease) family resemblance (in Type 1 resistant rickets) and present drugs (anticonvulsant osteomalacia).

Physical signs

Chief amongst the physical signs of metabolic bone disease are the appearance, particularly of the face and skull, the proportions of the skeleton, and the shape of the bones. When the subjects are seen as outpatients the gait, height and proportions can be easily checked. This is not so for patients first seen in bed in hospital where the medical attendant may not know how tall a patient is, whether or not he can walk, and if so, whether the gait is abnormal.

Appearance

(1) The face and skull
The appearance of the face and skull may give vital clues not only in inherited disease. Examples are the large vault in Paget's disease, the smooth skin and low hairline of the hypogonad male, the round wrinkled face of hypopituitarism, the elfin face of idiopathic hypercalcaemia, the prominent brow of Type 1 resistant rickets, the fair hair and ruddy complexion and glasses (due to lens dislocation) in homocystinuria, the anaemia and pigmentation of underlying renal disease (renal osteodystrophy), the coarse features, enlarged nose and lower jaw and widely-spaced teeth in acromegaly, and the 'round-faced' simplicity and cataracts of pseudohypoparathyroidism. Many more examples such as hypothyroidism and mucopolysaccharidosis are mentioned in the ensuing chapters. It is disappointing how little attention is paid to the facial appearance in diseases of the skeleton, apart from the description that it is sometimes 'odd'. Specific features of the face should also receive attention. This is particularly so for the eyes.

(2) The eye
Here one may see corneal calcification (*Figure 2.4*), arcus juvenilis (*Figure 7.2*), corneal clouding, cystine crystals (*Figure 2.5*), Kayser—Fleischer rings, dislocation of the lens (*Figure 2.6* and Chapter 8) and cataract. Calcification of the cornea is variously described, and it does not necessarily mean the same to the ophthalmologist as to the physician. Most commonly seen as a sign of present or previous hypercalcaemia, it is a crescentic shaped area of calcification on the medial and/or lateral borders of the limbus. It is to be distinguished from the senile arcus which tends to start at the top and bottom of the limbus before becoming complete. Statement about whether calcification is continuous with the sclerae are conflicting. Most often there seems to be a gap between the calcification and the scleral margin (*but see Figure 2.4*). Corneal calcification is detectable with the unaided eye or with a hand lens: its granular character should be confirmed with a slit lamp. Another sign of hypercalcaemia, less

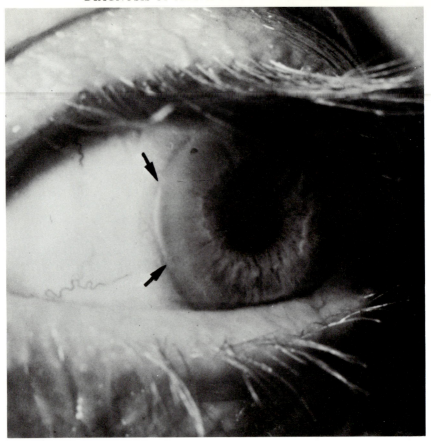

Figure 2.4. To show a white crescentic area of calcification at the lateral aspect of the cornea in a woman with hypercalcaemia

commonly seen is 'band keratopathy', in which there is a band of calcium across the front of the cornea. Conjunctival calcification also occurs.

Although a senile arcus has no particular value in the diagnosis of bone disease, a complete arcus may be seen in younger people with osteogenesis imperfecta (*Figure 7.2*). Affected persons may also have a non-circular cornea, in addition to the well recognized blue sclerae (which also occur occasionally in other connective-tissue disorders, in the extremes of age, and, it is said, sometimes in rickets). Keratoconus is an occasional feature of inherited disorders of connective tissue.

It is important to look closely at the eyes of a child with skeletal deformity or rickets. Corneal clouding (as in MPS IH and MPS IV) may be seen unaided, but cystine crystals (*Figure 2.5*) are not so easy to detect even with a hand lens; cystinosis should be suspected in a miserable dehydrated photophobic child with rickets and should be an indication for slit-lamp examination. The Kayser–Fleischer rings of Wilson's disease are characteristically of a brownish-gold colour and seen at the top and bottom of the cornea; easily seen with the aid of a torch against blue eyes, they may be very difficult to detect against brown eyes.

Dislocation of the lens is easy to confirm, and should be suspected in any patient with a Marfan-type stature. The shimmering of the unsupported iris (iridodonesis) is an important sign. Dislocation may occur upwards, downwards or laterally (*Figure 2.6*). Premature cataract occurs in hypoparathyroidism. In the very rare type of Ehlers–Danlos syndrome with hydroxylysine deficiency (Chapter 8), fragility of the globes of the eye, and microcornea may occur.

(3) Other facial features

Examination of the teeth should not be neglected, since dentine and bone matrix are chemically similar and disturbances of the skeleton might be expected to be associated with abnormality of the teeth. This association is not always present. Thus the teeth may be clinically normal in rickets, in juvenile osteoporosis, and in some patients with osteogenesis imperfecta. However, dental abnormalities are found in severe osteogenesis imperfecta, in hypoparathyroidism and in

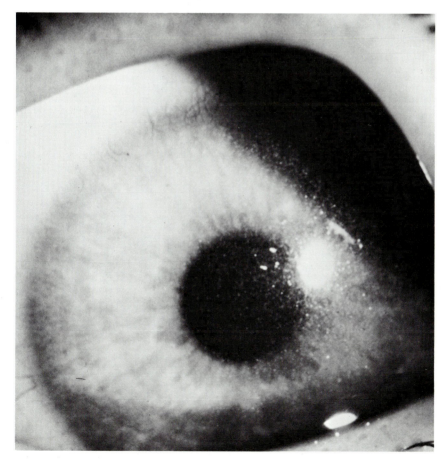

Figure 2.5. To show the 'snowstorm' appearance in the eye of a boy with cystinosis. The cystine crystals are in the cornea

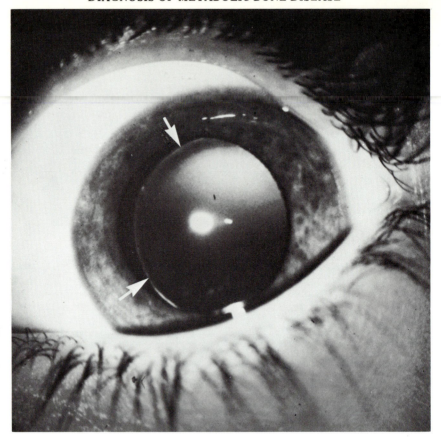

Figure 2.6. Lateral dislocation of the lens in a patient with Marfan's syndrome (left eye). Compare with the downward displacement in homocystinuria (Figure 8.5)

hypophosphatasia. It is diagnostically useful to distinguish between defects of dentine and enamel. In medullary carcinoma, multiple mucosal neuromas, together with a Marfanoid appearance, may provide clues to the diagnosis. Abnormal ears may give clues to chromosomal abnormalities and to congenital contractural arachnodactyly, and webbing of the neck suggests Turner's syndrome. The ears may also provide the first clues to alkaptonuria, with pigmentation of the cartilage (also in the nasal cartilage) and dark wax.

(4) The hands and feet
The fingers may be abnormally long and thin (arachnodactyly), or very mobile (Ehlers–Danlos syndrome); there may be abnormalities of the nails, such as moniliasis in hypoparathyroidism or absence as in the nail-patella syndrome. The hand may be short and wide, with splayed fingers (trident hand) as in some types of mucopolysaccharidosis, or with some short metacarpals (as in pseudo-hypoparathyroidism); or with accessory digits (Ellis–van Creveld syndrome).

There may be extensive subcutaneous calcification (scleroderma, hypo-parathyroidism). The thumbs may be short, suggesting myositis ossificans progressiva; in which case the monophalangic big toe is characteristic (Chapter 10). McKusick (1972) gives many more examples.

Stature and bodily proportions

There is often considerable ignorance about normal growth and bodily proportions (p. 35) and a reluctance either to measure these in patients or to relate them to established standards. Since all children grow, any orthopaedic assessment which fails to measure this growth is incomplete. Examples of abnormal growth are given in *Figure 2.3,* and in *Figure 9.6* and the causes of short stature are discussed on p. 66.

In adults, as in children, assessment of bodily proportions and their relation to each other, is important. The best known and most useful measurement is that of the upper to lower segment, with the dividing line at the top of the pubis symphysis. Whilst it is generally considered that the ratio of upper to lower segments is about 1, McKusick (1972) finds that it is about 0.93 *(see also* Chapter 8). In approximate terms, in the adult, the height is equal to span, and the upper segment is equal to the lower. Some variations from this are: racial — in some negroes the lower segment is relatively longer than the upper; endocrine — as in hypogonadism, where the upper to lower segment ratio is reduced, due to excessively long limbs; rickets — such as the hypophosphataemic vitamin-D-resistant type, with disproportionately short limbs; achondroplasia — the prototype of short-limbed dwarfism; and the inherited disorders of connective tissue — relatively long limbs in Marfan's syndrome and homocystinuria, short limbs in severe osteogenesis imperfecta.

The spine

Examination of the spine is included in the general assessment, but specific features are important. The commonest change is the thoracic kyphosis of osteoporosis. Scoliosis is an important sign, which may in some circumstances indicate an underlying connective-tissue disorder. The thoracolumbar gibbus is a particular (but not exclusive) feature of the mucopolysaccharidoses. A sharp angular deformity of the spine which may be dominantly inherited is seen in neurofibromatosis. Changes in the spine do not occur in isolation. Thus a severe scoliosis will produce a considerable deformity of the chest and alterations of the structure and function of the organs within it such as occasional cardio-respiratory failure. With the kyphosis of osteoporosis, especially when it occurs in the relatively young, the sternum becomes prominent and a transverse crease develops across the abdomen. The ribs may press on to the iliac crest.

Other features

Other features which provide clues to diagnosis of biochemical disorders of the skeleton include neurofibromata and pigmentation of various sorts (neuro-fibromatosis, polyostotic fibrous dysplasia), abnormal scars with poor healing, ('collagen' disorders), and bizarre asymmetry of bone (neurofibromatosis). Examination of other systems may demonstrate cardiac abnormality, such as aortic regurgitation or mitral incompetence (Marfan's syndrome), subvalvar aortic stenosis (idiopathic hypercalcaemia), evidence of venous thrombosis (homocystinuria), or splenomegaly (sarcoidosis). Other features will be mentioned in ensuing chapters, but it is appropriate here to discuss tetany.

Tetany   Spontaneous tetany is uncommon. It is associated with hypocalcaemia in rickets and osteomalacia and may also follow the alkalosis of overbreathing. It is fully described by Fourman and Royer (1968) and illustrated by Paterson (1974). Characteristically the hands go into the position of carpal spasm with adduction and flexion of the thumb into the palm of the hand and with adduction of the extended fingers. The fingers become slightly flexed at the metacarpo-phalangeal joints and the wrists also become flexed. Tetany may occur intermittently and be mistakenly regarded as psychological. In children with rickets more dramatic features such as laryngeal stridor can occur. Other features of hypocalcaemia include tingling of the fingers and around the mouth, and if the hypocalcaemia is of sudden onset — as after parathyroidectomy — fits or depression may occur. Prolonged hypocalcaemia in hypoparathyroidism may produce depression, confusion, psychosis and fits (Chapter 6).

Although spontaneous tetany is rare, there are two recognized tests for latent tetany. Chvostek's sign is elicited by tapping the branches of the facial nerve (with a finger or tendon hammer), as they spread out from within the parotid gland. This should be carried out in a systematic manner from above downwards and may produce a twitching of the appropriate muscles above the eyebrow, of the upper lid, of the corner of the nose and upper lip, of the lower lip and below this. This test is supposed to demonstrate increased excitability of the facial nerve. It may be positive with a normal plasma calcium (especially in young females) or negative with hypocalcaemia, but despite these reservations, it is a useful clinical indicator of a low plasma calcium. The test may also be positive with hypomagnesaemia.

Trousseau's sign is more troublesome to do and more painful for the patient. Since it is often done in the wrong way — which adds to the discomfort but not to the diagnosis — the procedure is described. The aim is to render the forearm ischaemic to facilitate carpopedal spasm. To this end a sphygmomanometer cuff is applied to the upper arm, inflated to obliterate the pulse and left on for not more than three minutes. In a positive test carpal spasm will slowly occur with opposition of the thumb as the first movement. The forearm remains painful with tingling for a few minutes when the cuff is removed, and relaxation of the carpopedal spasm is slow. A positive test has the same significance as a positive Chvostek's sign although a false positive is probably less frequent.

BIOCHEMISTRY

After the history and physical examination, the next most useful investigations are biochemical. These are well discussed by Paterson (1974) and in the multi-author book by Nordin (1976). Diagnosis is most often based on the plasma values of calcium, inorganic phosphate and alkaline phosphatase. Magnesium is not measured routinely but should not be completely ignored.

Plasma

Common errors arise from failure to recognize changes with age (allied to the disregard of physical growth), from the continued alteration of units, and from the overemphasis of the plasma calcium to the neglect of phosphate. Whilst the fasting plasma calcium remains virtually constant through childhood and early adult life, plasma phosphate declines with adolescence and the alkaline phosphatase increases during the phase of rapid growth. In later life, plasma phosphate increases in females, as does fasting plasma calcium. In all stages of life seasonal changes may occur, associated with vitamin-D lack in the winter.

Calcium

The normal level of fasting plasma calcium varies slightly from one laboratory to another and whether it is measured by atomic absorption spectrometry or with an auto-analyser. It is within the range (± 2SD) of 9.0 to 10.2 mg per 100 ml or 2.25 to 2.55 mmol per litre. Since nearly half of the circulating calcium is bound to protein and the remainder (except for a small complexed fraction) is ionized, the amount of protein in the sample analysed is important. The specific gravity (sp. gr.) of plasma is linearly related to its albumin content. Acceptable corrections to a constant protein level, of 4.0 g of albumin per 100 ml, or to a plasma specific gravity of 1.027, can be made (Berry *et al.*, 1973; Payne *et al.*, 1973). In practice the correction should be made either by the laboratory, or from the figures available. Thus in a patient with a recorded plasma calcium of 10.4 mg per 100 ml, and a plasma specific gravity of 1.025, we correct this to what the plasma calcium would be at a normal protein concentration with a sp. gr. of 1.027. This correction is to increase the observed plasma calcium by 0.25 mg per 100 ml for each unit difference in the 3rd decimal place of the sp. gr. (Dent, 1962). In this instance the corrected plasma calcium would then be 10.4 plus 0.25 × 2 = 10.9 mg. Since routine biochemical screening includes a measurement of albumin, correction based on this is of wider application. The formula of Payne and his colleagues (1973), where corrected calcium, mg % (mg per 100 ml) = measured calcium − albumin, g % (g per 100 ml) + 4 is appropriate*.

It is not routine to measure the level of ionized calcium. Although direct-reading electrodes have made this theoretically more easy, practical difficulties still exist. The calcium fractions in the plasma are discussed by Marshall (1976). He points out that abnormally high or low total calcium values may be taken to indicate similar changes in ionized calcium. Difficulties however arise with values near to the normal range. Since there appears to be individual differences in protein-binding of calcium (Pain *et al.*, 1975), corrections are most useful when applied to sequential measurements in the same individual.

The fasting plasma calcium is normal in osteoporosis, except rarely in acute generalized immobilization; it is also normal in Paget's disease without immobilization. It is high in primary hyperparathyroidism, vitamin D overdosage, sometimes in neoplasms (of various sorts), and in a number of other states which may include acromegaly and thyrotoxicosis (Chapter 6). It is often low in osteomalacia, especially in the absence of secondary hyperparathyroidism, but it can

*In SI units an approximate correction is 0.02 mmol/l for every 1 g/l change of albumin from 40 g/l.

be normal and this is expected in vitamin-D-resistant hypophosphataemic rickets and various other forms of renal-tubular rickets. It is low in hypoparathyroidism and pseudohypoparathyroidism. It is normal in the inherited disorders of connective tissue.

Phosphate

The plasma phosphate is higher in childhood than in adult life, but the cause for this is not known. Concentrations quoted by Robertson (1976) from older literature are, in mg per 100 ml, 3.7–8.5 (age 0–12 months), 3.6–5.9 (children) and 2.4–4.5 (adults). The changes in children are seen in more detail in *Figure 2.2*, taken from Round (1973), which gives a narrower range. The conversion factor from mg per 100 ml to mmol per litre is 0.32.

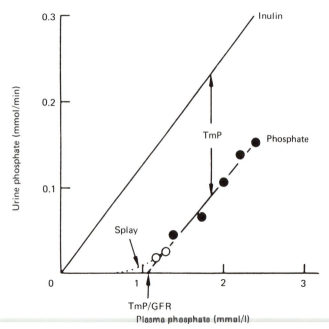

Figure 2.7. To demonstrate the relationship between plasma and urine phosphate. For explanation see text. (From Walton, Russell and Smith, 1975)

In children it is particularly important to know the normal ranges for age since the diagnosis of such disorders as hypophosphataemic rickets depends on this knowledge.

The fasting plasma phosphate is largely controlled by changes in renal phosphate excretion. If the plasma phosphate is artificially elevated by increasing the rate of infusion of phosphate there is a linear relationship between excretion rate and plasma levels, which is parallel to the GFR (*Figure 2.7*). The vertical distance between these straight lines represents the maximal reabsorption of phosphate called the TmP (maximum tubular reabsorption for phosphate). The line for phosphate excretion may be extrapolated to zero; the intercept on the plasma line is equal to the TmP/GFR since at the theoretical state where the urine excretion of phosphate is zero

$$UV = TmP, \text{ and GFR} = \frac{TmP}{P}$$

At the theoretical phosphate threshold TmP is also equal to filtered load, which is itself equal to GFR × phosphate concentration in the plasma at this threshold (P_{thres}). Thus

$$TmP = GFR \times P_{thres}, \text{ and the } P_{thres} = \frac{TmP}{GFR}$$

For further information the reader should consult Bijvoet (1977). Where the GFR is normal the plasma phosphate will vary with TmP, and those factors which alter TmP will alter plasma phosphate. The most potent of these is parathyroid hormone. Where there is renal-glomerular failure, the prime determinant of plasma phosphate is not tubular reabsorption but GFR (Robertson, 1976).

Thus the plasma phosphate is low in primary hyperparathyroidism and in hyperparathyroidism secondary to malabsorption. There is hypophosphataemia also in a number of renal-tubular disorders and in vitamin-D-resistant rickets (Chapter 4). A low plasma phosphate may also occur in aluminium hydroxide overdosage or during prolonged intravenous nutrition. Hyperphosphataemia occurs in hypoparathyroidism and in renal-glomerular failure and an inherited increase in TmP is the cause of the hyperphosphataemia of tumoral calcinosis.

Plasma phosphate is normal in osteoporosis and in Paget's disease (although a variation with disease activity has been described in this disorder (Chapter 5).

One should note that there has been much controversy about the various indices to be used in assessment of renal phosphate 'clearance' but following the work of Bijvoet (1969) and Bijvoet, Morgan and Fourman (1969) the use of TmP/GFR which takes into account filtered load is increasingly accepted. Nomograms for its derivation from simultaneous fasting values of P and creatinine in plasma and urine have been devised (Walton and Bijvoet, 1975). For practical purposes it is important to take as much notice of the plasma phosphate as that of calcium.

Alkaline phosphatase

The plasma alkaline phosphatase in the normal adult male is between 4 to 12 K.A. units per 100 ml (28–85 iu. per litre). It is slightly higher in males than females. In childhood it is increased above adult values and reaches a peak during the rapid growth of adolescence, with an upper limit of about 30 K.A. units per 100 ml. However there is likely to be considerable individual variation and one must always hesitate to attribute an elevated alkaline phosphatase to bone disease in a teenager unless it is markedly increased. The levels are closely related to growth and begin to fall earlier in girls than boys (Round, 1973; *Figure 2.1*). It is always important to check on the units used and the normal range for the particular laboratory.

Plasma alkaline phosphatase is normal (for age) in juvenile and adult osteoporosis, and in the inherited disorders of connective tissue. It is usually increased in osteomalacia due to vitamin D deficiency or malabsorption but not necessarily in osteomalacia due to renal-tubular disorders. It is increased in Paget's disease roughly in proportion to the radiological extent of the disease, but may be normal. In primary hyperparathyroidism it is usually normal, and as a general rule appears to be slightly increased in some disorders with excessive or abnormal bone formation such as Engelmann's disease and fibrous dysplasia. Finally there are those conditions which are named after the alteration in alkaline phosphatase, namely hypophosphatasia and hyperphosphatasia (Chapter 11), both of which are very rare.

An increase in the circulating alkaline phosphatase may be due to different isoenzymes from different sources, of which the bone and liver are most important. In normal persons the liver alkaline phosphatase remains virtually constant during growth. However an increase in plasma alkaline phosphatase does not necessarily mean bone disease. There are many ways of distinguishing the bone enzyme from enzymes from other sources, such as acrylamide gel electrophoresis or heat inactivation. An increase in the plasma level of 5-nucleotidase suggests liver disease (Nordin, 1976). It is also useful to measure the urine hydroxyproline since this correlates well with the increased alkaline phosphatase due to bone disease (p. 148) and is unaltered by liver disease. However, this does require an accurate 24-hour collection of urine on a low gelatin diet.

Magnesium

The normal range for plasma magnesium is 1.77 to 2.41 mg per 100 ml (0.75 to 1.0 mmol/l) and it does not change with age or sex. Hypomagnesaemia has several causes, of which steatorrhoea is important, where magnesium deficiency may be associated with (and sometimes causes) hypocalcaemia.

Other plasma measurements

Although the plasma calcium 'phosphate' and alkaline phosphatase are the main biochemical measurements used in bone disease, one should note other relevant changes which may be found on the routine multichannel biochemical screening, and also measurements which are potentially useful in research. In the former category there may be changes in the blood urea (low in malabsorption, high in renal glomerular failure), in sodium and potassium (low in cystinosis and renal-tubular acidosis); in CO_2 and chloride (transient hyperchloraemic acidosis in primary hyperparathyroidism and vitamin D deficiency, permanent in renal-tubular acidosis); and in plasma proteins (monoclonal increase in globulins in multiple myeloma). Other measurements which may be useful in bone research include those used for PTH, vitamin D metabolites and calcitonin. PTH assays differ according to their specificity and must be interpreted according to the normal values for the laboratory concerned, the prevailing level of plasma calcium and the GFR. The measurement of normal levels of calcitonin is controversial (*see* Chapter 12). Of the vitamin D metabolites, the measurement of 25 (OH) D has proved very useful in the study of vitamin D metabolites and osteomalacia (Chapter 4). Plasma non-protein-bound hydroxyproline, proline iminopeptidase (p. 150), and α_2HS-glycoprotein (p. 18) have also given some information about renal bone disease and Paget's disease.

Urine

Examination of the urine is often neglected in bone disease, partly because of the difficulty in obtaining 24-hour urine collections. A random specimen should be tested for glucose (renal glycosuria, Fanconi's syndrome) and protein (multiple myeloma, tubular proteinuria). The presence of a reducing substance which is not glucose should suggest alkaptonuria. If tubular abnormalities are suggested other functions should be examined — namely, the ability to concentrate the urine after overnight restriction of fluid and to acidify it after oral ammonium chloride. In selected cases (Chapter 4) amino acid chromatography is important.

A timed 2-hour fasting urine specimen, together with a mid-time fasting blood can be used to estimate TmP/GFR from creatinine and phosphate content (p. 51). A 24-hour urine should be collected for measurement of calcium and hydroxyproline; for hydroxyproline (Chapter 1) the patient should be on a low gelatin diet.

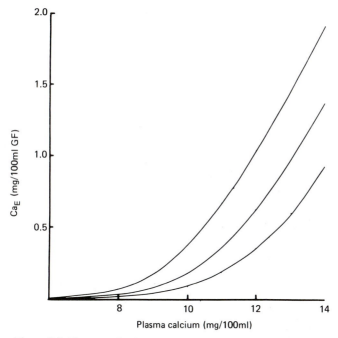

Figure 2.8 The normal relationship between the urine calcium and plasma calcium. This has been established by infusing normal persons to make them hypercalcaemic. In hypoparathyroidism points lie to the left of the curve and in hyperparathyroidism to the right. The lines enclose the normal range. (From Nordin, 1976)

Calcium

The relationship between calcium excretion (Ca_E) per 100 ml, glomerular filtrate (GF) and the plasma calcium has been established by the infusion of normal subjects with calcium (*Figure 2.8*). The renal handling of calcium in disease states can then be compared with the normal range.

The normal excretion rate of calcium is much debated, and an upper value of 400 mg (10 mmol) per day has been quoted. The output normally increases little with intake but this increase is more marked where there is excessive intestinal absorption of calcium in one of the types of idiopathic hypercalcuria. Nordin (1977) discusses the values in detail, and puts the 90th percentile for a Leeds population at 380 mg per day, noting the wide differences in different regions in the U.K. The lower limit of urine calcium will depend on the plasma calcium. At a plasma calcium of 2.25 mmol per litre and normal glomerular filtration rate, the corresponding excretion in the urine would be 144 mg per litre and in hypocalcaemic states it will be lower than this. Measurements of urine calcium are useful when they are excessively low, as in most forms of osteomalacia, or excessively high as in some hypercalcaemic states. Urine calcium excretion falls with renal-glomerular failure. One should note that since parathyroid hormone increases the renal reabsorption of calcium the urinary calcium excretion in patients with hyperparathyroidism is lower in relation to the plasma calcium than in normal subjects made equally hypercalcaemic by calcium infusions (*Figure 2.8*). Nevertheless, because of the hypercalcaemia, the urine calcium may be abnormally high in primary hyperparathyroidism, and also in situations of excessive bone resorption, for example, leukaemia, myeloma, secondary carcinoma and immobilization. In some hyperostoses, particularly marble bone disease, the urine calcium may be virtually undetectable (Chapter 11). These points are dealt with in detail by Robertson (1976).

Creatinine

Because of the relative constancy of individual daily urine creatinine excretion, it is widely used as a standard upon which to base the excretion of other substances, particularly where 24-hour urine collections are impossible or their accuracy is in doubt. However the normal range of urine creatinine in the adult is wide, from between 1 and 2 g per day, being higher in men. Comparison with creatinine excretion is most useful for random or consecutive urine collections.

Hydroxyproline

Since simple and specific methods capable of automation are available for hydroxyproline estimation, its measurement in urine is a useful guide to the diagnosis of metabolic bone disease.

Hydroxyproline is derived almost exclusively from collagen (Chapter 11) and a major part of the urinary hydroxyproline comes from bone collagen. It is usual to measure total hydroxyproline (THP) in a 24-hour urine, although the free, small peptide and large peptide (non-dialysable) fractions can be measured separately. Since hydroxyproline is derived from collagen its excretion in the urine is increased if the subject eats collagen, either native or denatured (gelatin). Therefore the patient should be on a low gelatin diet (which excludes substances like ice cream, jellies and mousses); a completely vegetarian diet is not usually necessary. The normal 24-hour values may be expressed as mg or mg per m^2, the surface area being derived from height and weight. The approximate daily value for adults is 15 to 50 mg (66–220 μmol), and the normal surface area about 1.7 m². Derivation of surface area from height and weight is not reliable in severe deformity. The ratio of hydroxyproline to creatinine in fasting urine is also used as an index of bone resorption. Before adult life THP excretion varies widely in proportion to the rate of linear growth, with a rapid and considerable increase in adolescence, earlier in females than males. Therefore at present one can only interpret values for THP excretion with any reliability in adults. Nevertheless increasing information is becoming available on THP excretion during growth,

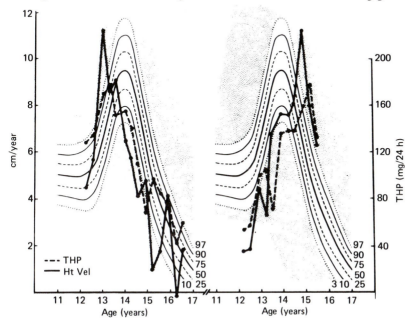

Figure 2.9 The relationship between height velocity and total hydroxyproline (THP) excretion. The examples given show that height velocity and THP correlate very well whether maximum growth rate is early or late compared with normal. (From Clark, 1977. Copyright by Academic Press Inc. (London) Ltd.)

and there is a close correlation between the percentiles for height velocity and hydroxyproline excretion (*Figure 2.9*; Clark, 1977).

In general THP excretion reflects bone collagen turnover and its breakdown, and the greatest increases are found in bone disease. It correlates well with the alkaline phosphatase, except under certain temporary circumstances, such as after parathyroidectomy (*Figure 6.6*). This correlation follows from the normal linkage between bone formation and resorption rate (Chapter 1). THP excretion is particularly increased in Paget's disease, and moderately increased in osteomalacia according to the degree of secondary hyperparathyroidism, and in primary hyperparathyroidism with radiological evidence of bone disease. It is within the normal range in osteoporosis although sometimes towards its upper limits. It is not significantly altered in osteogenesis imperfecta when compared with age-matched controls, but it is difficult to interpret the significance of hydroxyproline excretion in gross deformity where the skeleton is very abnormal (Chapter 7). The lowest values are found in myxoedema and hypopituitary dwarfism.

Phosphate

There is little point in measuring urine 24-hour phosphate excretion on its own; it varies widely with phosphate intake and needs to be related to the plasma levels so that reabsorption can be measured.

RADIOLOGY

Radiology plays a central role in the diagnosis of metabolic disorders of the skeleton, especially if no biochemical abnormalities can be readily demonstrated. It is not necessary to list the radiological appearance of all such diseases, since details are given in the appropriate chapters, but it is useful to make some simple points. Osteoporosis is the commonest metabolic disease of bone and it can be identified radiologically by the presence of structural failure. Apart from obvious fractures of the femoral neck, the forearm or other long bones, fractures also occur commonly in the vertebrae. These produce irregular localized vertebral wedging and asymmetrical biconcavity of the vertebral bodies, an appearance which is usually distinct from the generalized symmetrical biconcave changes in the osteomalacic spine. It is confusing to apply the term 'codfish' or 'fish-like' to the irregular vertebral collapse of osteoporosis. Radiographs of the spine will not necessarily help to say whether a person with osteoporosis also has osteomalacia, but where x-rays of the rest of the body do show diagnostic features of osteomalacia, that is, Looser's zones, there is no difficulty.

Paget's disease may show almost any feature (Chapter 5) but characteristically the affected bone is larger than it ought to be and, where comparison is possible, larger than its fellow.

The hallmark of parathyroid overactivity is the presence of subperiosteal erosions which occur typically in the phalanges. Localized areas of osteitis fibrosa may cause difficulty. The features (some of which are bizarre) of osteogenesis imperfecta, the mucopolysaccharidoses, and the hyperostoses, are dealt with in ensuing chapters. The clinician will learn more by looking at the radiographs rather than at the report.

FURTHER INVESTIGATIONS

Use of the information in the preceding pages will establish the diagnosis in the majority of patients. However additional studies may be necessary. These include bone biopsy, mineral balance, isotope uptake and skeletal scanning.

Bone biopsy

The bone is best taken by a transiliac trephine under a local anaesthetic. It is usual also to give the patient intravenous diazepam (Valium) which makes the procedure virtually painless and produces amnesia for the event. The biopsy obtained is a core of trabecular bone bounded by the inner and outer cortex.

Some bone should be prepared without decalcification, since the prior removal of mineral makes it very difficult to assess the amount of osteoid. Osteoporosis cannot easily be diagnosed on bone biopsy unless it is severe; the hallmark of osteomalacia is an increase of osteoid (either in thickness or in bone surfaces covered by it or both); Paget's disease is recognizable by its excess cellular activity with evidence of active bone resorption and formation, marrow fibrosis and mosaic lines. Interpretation of the histological changes in the rarer bone diseases, such as osteogenesis imperfecta, is open to much discussion.

Unless there is a clearly localized lesion, bone biopsies should always be taken from the same site by the same technique. This is usually the ilium because considerable information of normal bone from this area is available. The disadvantage of always using the same site is that some changes in the skeleton are very localized. Thus an iliac bone biopsy may be quite normal in a patient with obvious radiological Paget's disease elsewhere, and such a biopsy may also appear normal in a patient with clinical osteoporosis. Most investigators use the iliac crest, and concentrate on trabecular bone, but some biopsy ribs and look at cortical bone. Microradiography has been much used but it may be misleading since it gives no information about the cells and their activity. Most bone pathologists recommend (although they may not practise) that the patient is given tetracycline (Jowsey, 1977) before biopsy. This will 'label' the bone, since the tetracycline is incorporated into the mineralizing front. Two tetracycline labels separated by a given time will enable one to estimate the rate of bone formation.

These comments about bone biopsy refer to its use in diagnosis. Increasingly, quantitative measurements are made on the surfaces of appropriately prepared bone sections to obtain an estimate of the amount of bone undergoing formation and resorption. Quantitative measurements on normal bone from the same site differ widely. Such quantitative measurements are used mainly in research either to compare with some biochemical changes, such as parathyroid hormone level or to estimate the effect of treatment such as vitamin D metabolites. In view of the wide variation in normal bone, only considerable and consistent changes in quantitative measurements can be considered to be of significance.

Balance studies

The aim of a balance study is to measure the overall external balance of the mineral concerned. This naturally requires the accurate measurement of intake, but also, with elements such as calcium which are mainly excreted in the stools (in comparison with nitrogen), it requires an accurate measure of faecal output with methods of correction for faecal loss. It is necessary to allow the patient to 'equilibrate' to a constant intake, which should be as near to the out-patient intake as possible. Since there are day-to-day variations in recorded intake and output, one established way of getting an average value which minimizes errors is to do a long balance, but this consumes a lot of the patient's and biochemist's time. Certain refinements have therefore been added to make it possible to obtain significant results in a relatively short time; thus it is now possible (after equilibration) to do an accurate balance during two 4-day periods marked with external markers, by continuously feeding a non-absorbable and non-metabolized marker to detect faecal loss. Markers used include barium sulphate and copper thiocyanate (which seems very satisfactory). To use an example, if 200 mg of copper thiocyanate is given per day and less than 800 mg is recovered over four days (say 600 mg), then the recovery is less than one would expect. This means that the faecal calcium does not represent the whole of the 4-day excretion, and

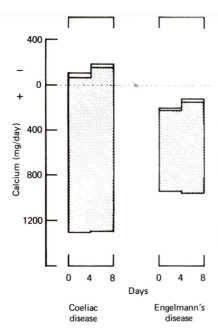

Figure 2.10. To demonstrate different external balances in two patients with low urine calcium excretion. Intake is plotted down from 0, and output upwards from the bottom of intake. The shaded area represents faecal calcium and the white blocks urine calcium. In the patient with coeliac disease faecal output is more than intake and there is a negative calcium balance. The patient with Engelmann's disease (Chapter 11) is in positive calcium balance, absorbing approximately 200 mg daily, despite excessive calcium in the skeleton

a correction of 4/3 would be therefore applied to the recorded figure. In fact the overall recovery over two 4-day balance periods is often very near 100 per cent. A critical appraisal of the balance technique is given by Lentner *et al.*, (1975). The external balance of phosphorus and magnesium is often measured with that of calcium, since they are both concerned with biochemical events in bone but, far more than calcium, they are found in muscle, red cells and other tissues.

Balance studies can be useful in both diagnosis and treatment, but should not be used merely to provide additional time for other investigations or for thought.

A useful example in diagnosis is provided by the patient with a very low urine calcium, which could be either due to failure of calcium absorption (as in osteomalacia) or to avidity of the bone for all available calcium (as in one of the hyperostoses). These different balances are compared in *Figure 2.10*.

There are times when one wishes to know whether a treatment is having the desired effect. Thus a given treatment might decrease the urine calcium in a patient with osteoporosis. This could indicate either an increased uptake of calcium by the skeleton (with a positive calcium balance), or it could be a consequence of a defect in calcium absorption through the gut (with a negative calcium balance). This point could be settled by a quick balance study. Further examples are provided by studies on the treatment of Paget's disease and the effectiveness of various vitamin D metabolites in renal osteodystrophy.

A balance study may be combined with the administration of the isotope ^{47}Ca which then gives information for the calculation of true calcium absorption, endogenous faecal calcium, and mineralization and resorption rate of bone (Nordin, 1976).

Isotopes

It would not be profitable to give a detailed account of the use of radioactive isotopes in the investigation of metabolic bone disease, for which Nordin (1976) should be consulted. The following is an outline of their uses.

Bone scanning

Various agents which concentrate specifically in bone can be used to demonstrate areas both of physiological overactivity (such as the growing ends of long bones) and pathological overactivity (such as tumours and Paget's disease). The exact physiological basis for their action is not understood but this does not prevent their increasing and widespread application (Lentle *et al.*, 1976).

All agents which are used for bone scanning are radioactive. A suitable scanning agent must not cause hazardous radiation by having a long biological half-life, or be impossible to use because of its very short physical half-life. For these reasons various isotopes of calcium, strontium and fluorine had only temporary use, and all modern scanning agents make use of the ^{99m} technetium label attached to the substances such as polyphosphate, pyrophosphate or phosphonate, of which the latter seems the most promising.

It is not clear why these isotopes concentrate in particular areas. Certainly they accumulate preferentially where new bone is being formed and in these areas there is considerable osteoblastic activity, with formation of new collagen and with subsequent mineralization. Increased vascularity will cause the isotope to arrive at the bone site more quickly but additional factors may contribute to its fixation at this site. Although technetium-labelled agents have proved particularly useful in bone disease, uptake does occur in other tissues, for instance infarcted cardiac muscle.

Bone scanning has found particular application in the detection of tumours not seen by conventional radiology. Localized abnormalities may be seen in a number of other conditions, such as Paget's disease and osteomyelitis. Generalized bone disease, such as rickets or hyperparathyroidism may also provide abnormal scans, but they do not necessarily improve diagnostic accuracy.

Lavender (1976) gives good examples of normal and abnormal bone scans, some obtained with ^{99m} technetium-labelled EHDP. He points out that bone scans differ from conventional radiographs, especially in young people, in showing increased uptake in the areas of the epiphyseal plates where there is little calcium. At all ages there is a greater concentration of uptake relative to radiographic density at the ends of the long bones, round joints, in the facial bones and at junctions with cartilage. In the detection of secondary bone tumours, comparison of scanning and total urine hydroxyproline excretion suggests that both can be abnormal when conventional radiography shows no change. It seems

TABLE 2.I
Features of some metabolic bone diseases

	Clinical (commonest symptoms)	With abnormal 'bone' biochemistry					Radiological	Comments
		Biochemical						
		Plasma			Urine			
		Ca	P	Pase	Ca	THP		
Osteomalacia (and rickets)	Bone pain Proximal myopathy	N or L*	L	N or H	L	N or H	Looser's zones Wide growth plate Cupped metaphyses	Plasma P high in renal failure
Paget's disease	Pain Deformity	N	N	H	N	H	Any appearance Bones often large	Plasma Ca high if immobilized
Hyperparathyroidism (with bone disease)	Pain Deformity Symptoms of hypercalcaemia	H	L	H	H	H	Subperiosteal erosions 'Cysts' in bone	Pase and THP normal without clinical bone disease
Hyperostoses Marble bones disease (severe)	Anaemia Nerve compression	N	N	H (acid)	L	N	Dense bones No remodelling	Acid Pase increases in marble bones disease
Engelmann's disease	Difficulty in walking	N or L	N or H	H	L	H	Diaphyseal hyperostosis	P may be high
Hypophosphatasia	Malaise Deformity in children; fractures in adults	N or H	N	L	N	N or L	Like rickets	Phosphoethanolamine in urine
Hyperphosphatasia	Deformity Large head Bent limbs in childhood	N	N	H	?N	H	Deformity (like Paget's)	Probable recessive inheritance Usually no mosaic lines in bones

	Ectopic mineral	N	H	N	?	?	Ectopic mineral over large joints	No renal failure Recessive inheritance
Tumoral calcinosis	Ectopic mineral	N					Ectopic mineral over large joints	No renal failure Recessive inheritance
						With normal 'bone' biochemistry'		
Osteoporosis	Loss of height Fractures				?		Irregular vertebral collapse	Urine Ca increased in acute osteoporosis
Marfan's syndrome	Long and thin Dislocated lenses						Long thin bones Scoliosis	Dominant family history
Homocystinuria	Dislocated lenses Mentally backward						Osteoporosis of vertebrae Large femoral heads	Recessive inheritance
		Homocystine in urine						
Osteogenesis imperfecta	Multiple fractures Blue sclerae						Thin deformed bones with fractures	Often dominant family history according to type
Alkaptonuria	Dark urine Premature arthritis						Calcified intervertebral discs	Recessive inheritance
		Homogentisic acid in urine						
Mucopolysaccharidoses	Growth failure Thoracolumbar kyphosis Intelligence reduced (depends on type)						Deformity of chest Trident hands Dysostosis multiplex	Variable, according to type
		Mucopolysaccharides in urine						
Myositis ossificans progressiva	Pain and swelling spinal muscles						Multiple bars of ectopic bone	Monophalangic big toe

This list does not include ectopic mineralization (due to many causes) where the bones appear normal. Detailed biochemical changes are dealt with in the appropriate chapters.

* N = Normal; H = High; L = Low.

TABLE 2.II

Scheme for the diagnosis of metabolic bone disease

Age	Main presenting symptom	Most likely diagnosis	Frequency	Exclude
Over 50 years	Pain in the back Loss of height Fracture	Osteoporosis Commonest in women	Common	Myeloma (especially in men) Secondary deposits Coexistent osteomalacia
	Deformity of long bones Pain in hips and pelvis	Paget's disease	Common	Osteomalacia Secondary deposits
	Bone pain Difficulty in walking	Osteomalacia	Uncommon, especially in the adult	Carcinoma Polymyalgia rheumatica
	Bone pain and deformity Thirst, nocturia, depression, vomiting, constipation	Osteitis fibrosa cystica	Rare	Carcinoma Myeloma
20–50 years	Loss of height Bone pain	Probably secondary deposits, or myeloma	Rare	Osteomalacia 'Idiopathic' osteoporosis
	Muscle weakness Loss of height	Osteomalacia	Rare	Late muscular dystrophy Neoplastic neuromyopathy

Age	Clinical features	Common cause	Frequency	Other causes
0–20 years	Bowing of bones Deformity Weakness	'Nutritional rickets'	Common in immigrants in Northern cities	Other causes of rickets (Chapter 4) Hypophosphatasia
	Multiple fractures Bruising	In infants, inflicted by parents. 'Battered baby'.	Not uncommon	Osteogenesis Imperfecta
	Bone pain Ill health	Leukemia	Uncommon	Osteomyelitis Osteomalacia
	Pain in back Difficulty in walking Pain in ankles Less rapid growth	Juvenile osteoporosis	Rare	Leukemia
	Failure to grow (dwarfism)	Many causes (Tables 2.III–2.VI)	Common	Particularly hypothyroidism and Turner's syndrome
	Excessive or disproportionate growth	Several causes	Less common	Particularly pituitary tumour Homocystinuria Hypogonadism & chromosomal abnormalities
	Fracture and deformity at birth	Severe osteogenesis imperfecta	Uncommon	Hypophosphatasia Achondrogenesis Thanatophoric dwarfism

likely that computerized axial tomography (EMI scanning) will in future provide a very good method for the detection of metastases, as well as for studying more generalized bone diseases (Kreel, 1976).

Calcium absorption

Measurement of the intestinal absorption of calcium may be obtained by comparison of the radioactivity in the blood after simultaneous oral and intravenous administration of different isotopes of calcium. Alternatively if a whole body counter is available absorption can be assessed by measurement of the whole body retention of calcium, with due correction for urine excretion. The conditions for all tests of calcium absorption must be standardized, since the results obtained depend on many factors, particularly the amount of non-isotope calcium incorporated as a 'carrier' for the isotope or the dietary calcium. In healthy persons there is a wide range of values for calcium absorption. One of the most useful applications of whole body counting has been to detect changes in individuals undergoing treatment, for instance with 1α-hydroxylated vitamin D metabolites in renal osteodystrophy.

Bone turnover

The activity of the skeleton may be assessed by measuring the disappearance rate of administered radioactive calcium into bone, taking into account faecal and urinary loss. This form of kinetic analysis is well summarized by Jowsey (1977), who points out that the results obtained are empirical and not necessarily related to actual events in bone. In particular the 'bone formation rate' or accretion rate' or 'V_o^+' is not necessarily related to the real rate of deposition of newly mineralized bone tissue. Such measurements are non-invasive and are useful provided their limitations are understood.

Mineral content

The mineral content of the body or part of it may be measured either by photon absorptiometry or neutron activation analysis. These methods have been particularly applied to osteoporosis and are dealt with briefly in Chapter 3.

DIAGNOSIS

Although the identification of the main bone disorders described in the subsequent chapters should not provide difficulty, a summary of their main distinguishing features (Table 2.I) may be useful. The disorders are not arranged in order of frequency but according to whether or not the biochemical findings referable to the skeleton are abnormal. These do not include those biochemical abnormalities produced by the inborn errors in alkaptonuria, homocystinuria and the mucopolysaccharidoses.

64

TABLE 2.III

Suggested scheme for the diagnosis of short stature

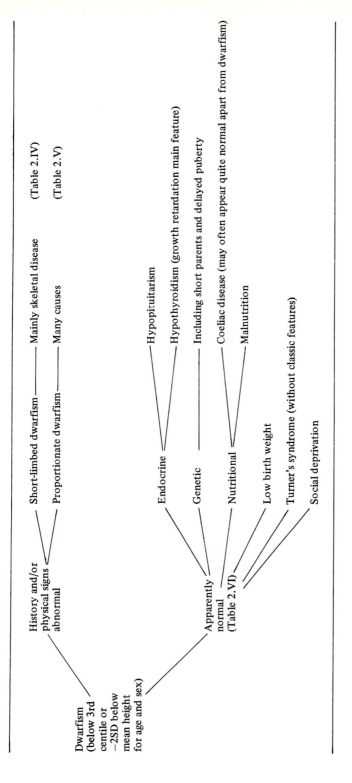

It can also be useful to have some sort of scheme to lead one to the correct diagnosis. Many schemes could be arranged, but the present one (Table 2.II) is based on the age of the patient when first seen. Age is an important point, since osteoporosis and Paget's disease are particularly common in later life, and disorders of growth are restricted to childhood and adolescence, although their effects may be lifelong. The reader may prefer to construct his own scheme from the information in this book, but it is important to have some method behind the diagnosis, if only to avoid missing treatable disease (such as osteomalacia or juvenile hypothyroidism).

In some patients the main clinical problem is abnormal height or disproportion. The most important of these is short stature.

Short stature

A child (or adult) is considered short if the height is below the 3rd centile, which is approximately 2 standard deviations below the mean height of the age and sex-matched population. Investigation of short stature has become very complex and in order to avoid unnecessary investigation whilst detecting remedial causes, the clinician should have some simple means of classification, suggested in Tables

TABLE 2.IV
Some causes of short-limbed dwarfism. (*See also* McKusick, 1972)

Condition	Genetics	Clinical features
Achondroplasia	Dominant	Very short limbs
Conradi's syndrome	Recessive	Punctate calcification of developing cartilage, mental retardation, high mortality
Diastrophic dwarfism	Recessive	Club feet, ossified ears, hitch-hiker thumb, scoliosis
Thanatophoric dwarfism	?	Short limbs, small chest, early death
Ellis—van Creveld syndrome	Recessive	Additional digits, congenital heart disease
Achondrogenesis	Recessive	Die in perinatal period. Very little ossification of limbs, vertebrae or pelvis
Severe osteogenesis imperfecta	? Sporadic Recessive	Multiple fractures, abnormal dentine (Chapter 7)
Cartilage-hair hypoplasia	Recessive	Very short. Sparse hair
Hypophosphatasia	Recessive	Neonatal forms most severe; phosphoethanolamine in urine
Hypophosphataemic rickets	X-linked	Low phosphate (Chapter 4)
Metaphyseal dysostosis	Dominant	May resemble severe rickets

DIAGNOSIS

2.III–V, and modified from Raiti (1969) and Parkin (1975). The history and examination of the short child may suggest severe or chronic illness, and the dwarfism may be proportionate or disproportionate. The commonest type of disproportionate dwarfism is short-limbed dwarfism and this is generally due to inherited skeletal disease (Table 2.IV). Disproportionate short stature is dealt with at length by Bailey (1973). Proportionate short stature (or dwarfism) with

TABLE 2.V
Some causes of proportionate short stature*

(1)	Genetic	Familial short stature
		Delayed puberty
(2)	Endocrine	Growth hormone deficiency
		Hypothyroidism**
		Corticosteroid excess (endogenous or exogenous)
(3)	Metabolic	Renal disease
		Liver disease
		Storage diseases (including mucopolysaccharidoses)
		Diabetes
(4)	Chronic disease	Congenital heart disease
		Fibrocystic disease
(5)	Nutritional	Coeliac disease
		Starvation
(6)	Intra-uterine	Low birth-weight dwarfism†
(7)	Chromosomal	Turner's syndrome (For others see Smith, 1976)
(8)	Social	Emotional deprivation

*   Most patients in groups 1–5 are likely to have abnormal history or physical signs.
†   The causes of low birth-weight dwarfism are multiple
** Hypothyroidism may produce short limbed dwarfism

evidence of organic illness may have many causes (Table 2.V). However a significant percentage of short children appear otherwise normal (although this percentage falls in those who are very short – further than 3 SD below the mean height) and the causes of this are relatively few (Table 2.VI). It is very important to note that in this group there are some patients with hypothyroidism with minimal physical signs (*Figure 2.3*), with coeliac disease, and with Turner's syndrome (without the classic features). These disorders should always be considered and if necessary excluded. Clearly the disorders in Tables 2.V and 2.VI overlap. It is important that if any doubt exists, children with short stature should be fully investigated by a paediatrician.

Since social deprivation may be the commonest pathological cause of short stature in this country, screening of children with short stature should take this into account. Parkin (1975) suggests that in this screening it should be possible to identify three groups, namely those in whom the possible cause is social, those who are quite fit and are short for genetic reasons, and those who require further investigation, either because of organic disease, very short stature (more than 3SD below the mean), or a combination of short stature (below the third percentile) and slow rate of growth. If after initial examination, including assessment of puberty, visual fields, radiological bone age and skull x-ray, there

TABLE 2.VI

Causes and features of dwarfism in apparently normal children

Hypopituitarism	Birth weight usually normal. Growth retardation within first 2 years of life. Progressive fall below 3rd centile. Often fat. No organic cause found for growth hormone deficiency in about 50%. Gonadotrophins may be impaired at puberty. In adult life may have features of hypogonadism, non-fusion of epiphyses and osteoporosis.
Hypothyroidism	Growth rate very slow. Bone age invariably retarded. Infantile upper segment/lower segment ratio. Characteristic stippled epiphyses.
Genetic	Familial short stature, with short parents. Delayed development and puberty may also be genetically determined — adult height may be normal.
Nutritional	Most difficult diagnostic problem is mild coeliac disease. Presents with short stature at any age. Anaemia, muscle wasting, diarrhoea (sometimes). Malnutrition is commonest cause of short stature world-wide. Mild malnutrition in social deprivation.
Low birth weight	Low birth weight for dates. Multiple causes. Intra-uterine effects, chromosomal disorders and others.
Turner's syndrome	Relatively common. 1 in 5000 female births. Webbing of neck, increased carrying angle elbow, shield-like chest, congenital heart disease, short stature, failure of sexual development. Short stature may be only feature in children.
Emotional deprivation	Common cause of failure to thrive and dwarfism. Temporary hypothalamic and pituitary disturbance. Responds to secure environment.

are no clues, further investigation must be done to exclude hypothyroidism, hypopituitarism, coeliac disease and Turner's syndrome. It is also important to exclude renal disease, and there are not always clinical clues to this.

It is essential, both for diagnosis and treatment of abnormal stature, always to compare radiological bone age (using Greulich and Pyle or Tanner and Whitehouse standards) with chronological age.

Tall stature

Tall stature is not as frequent a problem as short stature. Its commonest cause is genetic, and eventual height is often 'normal' when allowance is made for the height of the parents. Advanced development (the opposite of delayed puberty) does not result in an increase in adult height. Excess height may occur in chromosomal abnormalities with the XXY anomaly, and in cerebral gigantism, and can be a very troublesome feature of Marfan's syndrome, especially in girls (Chapter 8). In practice, difficulties arise in deciding whether an excessively tall adolescent has pituitary overactivity, and in predicting eventual height.

REFERENCES

Bailey, J. A. II. (1973). *Disproportionate Short Stature.* London: W. B. Saunders

Berry, E. M., Gupta, M. M., Turner, S. J. and Burns, R. R. (1973). Variations in plasma calcium with induced changes in plasma specific gravity, total protein and albumin. *Br. med. J.* **4**, 640–643

Bijvoet, O. L. M. (1969). Relation of plasma phosphate concentration to renal tubular reabsorption of phosphate. *Clin. Sci.* **37**, 23–36

Bijvoet, O. L. M. (1977). Kidney function in calcium and phosphate metabolism. In *Metabolic Bone Disease,* Eds. L. V. Avioli and S. M. Krane. Vol. 1. pp. 49–140. London and New York: Academic Press

Bijvoet, O. L. M., Morgan, D. B. and Fourman, P. (1969). The assessment of phosphate reabsorption. *Clinica chim. Acta* **26**, 15–24

Clark, S. (1977). Longitudinal growth studies in normal and scoliotic children. In *Scoliosis. Proceedings of a 5th Symposium, 1976.* Ed. P. A. Zorab. London and New York: Academic Press

Dent, C. E. (1962). Some problems of hyperparathyroidism. *Br. med. J.* **2**, 1419–1425, 1495–1500

Fourman, P. and Royer, P. (1968). *Calcium Metabolism and the Bone.* 2nd edn. Oxford: Blackwell Scientific Publications

Jowsey, J. (1977). *Metabolic Diseases of Bone.* Vol. I. London: W. B. Saunders

Kreel, L. (1976). The EMI general purpose scanner. A clinical report. In *Advanced Medicine.* Ed. D. K. Peters. Vol. 12 pp. 213–224

Lacey, K. A. and Parkin, J. M. (1974). Causes of short stature. *Lancet,* **2**, 42–45

Lavender, P. (1976). Radioisotope bone scanning. In *Advanced Medicine.* Ed. D. K. Peters. Vol. 12 pp. 213–224

Lentle, B. C., Russell, A. S., Percy, J. S., Scott, J. R. and Jackson, F. I. (1976). Bone scintiscanning updated. *Ann. intern. Med.* **84**, 297–303

Lentner, C., Lauffenberger, T., Guncaga, J. Dambacher, M. A. and Haas, H. G. (1975). The metabolic balance technique: a critical appraisal. *Metabolism* **24**, 461–471

Marshall, D. H. (1976). Calcium and phosphate kinetics. In *Calcium Phosphate and Magnesium Metabolism.* Ed. B. E. C. Nordin. pp. 257–297. London and Edinburgh: Churchill Livingstone

McKusick, V. A. (1972). *Heritable Disorders of Connective Tissue.* 4th edn. St. Louis: C. V. Mosby, Co

Nordin, B. E. C. (1976). *Calcium Phosphate and Magnesium Metabolism.* London and Edinburgh: Churchill Livingstone

Nordin, B. E. C. (1977). Hypercalciuria. *Clin. Sci. Molec. Med.,* **52**, 1–8

Pain, R. W., Rowland, K. M., Phillips, P. J. and Duncan, B. McL. (1975). Current 'corrected calcium' concept challenged. *Br. med. J.* **4**, 617–619

Parkin, J. M. (1975). Disorders of growth. In *Medicine. 2nd Series. 9. Endocrine Diseases.* Part 3. 409–412

Paterson, C. R. (1974). *Metabolic Disorders of Bone.* Oxford: Blackwell Scientific Publications

Payne, R. B., Little, A. J., Williams, R. B. and Milner, J. R. (1973). Interpretation of serum calcium in patients with abnormal serum proteins. *Br. med. J.* **4**, 643–646

Raiti, S. (1969). The short child: clinical evaluation and management. *Br. J. Hosp. Med.* **2**, 1640–1643

Robertson, W. G. (1976). Plasma phosphate homeostasis. In *Calcium Phosphate and Magnesium Metabolism.* Ed. B. E. C. Nordin. pp. 217–229. London and Edinburgh: Churchill Livingstone

REFERENCES

Round, J. M. (1973). Plasma calcium, magnesium, phosphorus and alkaline phosphatase levels in normal British schoolchildren. *Br. med. J.* **3,** 137–140

Smith, D. W. (1976). Recognizable patterns of human malformation. Genetic, Embryologic and Clinical Aspects. 2nd ed. Vol. VII. In *Major Problems in Clinical Paediatrics.* Ed. Alexander J. Schaffer. London: W. B. Saunders Co.

Tanner, J. M. (1973). *Growth at Adolescence.* 2nd edn. Oxford: Blackwell Scientific Publications

Vaughan, J. (1975). *The Physiology of Bone.* 2nd edn. Oxford: Clarendon Press

Walton, R. J. and Bijvoet, O. L. M. (1975). Nomogram for derivation of renal threshold phosphate concentration. *Lancet* **2,** 309–310

Walton, R. J., Russell, R. G. G. and Smith, R. (1975). Changes in the renal and extrarenal handling of phosphate induced by disodium etidronate (EHDP) in man. *Clin. Sci. Molec. Med.* **49,** 45–56

3

Osteoporosis

INTRODUCTION

Osteoporotic bone is normal in composition but reduced in amount (Albright and Reifenstein, 1948). This reduction must result from an imbalance of bone resorption over formation. Such an imbalance occurs with increasing age (Exton-Smith, 1976), and osteoporotic bones are therefore most commonly seen in the elderly. Whether or not this change is to be regarded as physiological or pathological is probably irrelevant, since one is merely an advanced form of the other. What is clinically important is its relation to structural failure, that is, to fracture, most often of the vertebrae, the femur or the forearm.

In view of these apparently simple statements, the reader may wonder why so much has been written on osteoporosis with so little agreement, why the cause of age-related bone loss is not known, and why its relation to fracture rate remains confused.

Since bone formation rate appears to decline from about the age of 40 years, whilst resorption continues and may increase, this will inevitably lead to a reduction of bone mass. In women this reduction is far greater than in men and follows the menopause. The apparently obvious conclusion that excessive bone loss in women compared with men is due to the postmenopausal decline in female hormones (which clearly does not occur in men) has been difficult to demonstrate. Likewise, since so many biological events occur with age, it has been difficult to identify the causal, rather than the coincidental, factors in the thin bones of old age.

Since bones become thinner with age and fracture rates increase, it is natural to consider that one is the cause of the other. But thinning of bones is only a contributory cause of fracture since there are other important factors which alter with age. It is well known, for instance, that elderly people fall more often than their younger relatives. This might be expected to have a greater effect on long bone fracture than vertebral fracture. Other factors such as drugs may be important. Thus Muckle (1976) showed that in 200 femoral neck fractures, 60 per cent occurred without a significant history of injury and in this group about

71

70 per cent of subjects were on long-term medication which would affect bone density. However, no figures were given for an age-matched group without fractures. If we take groups of patients at different ages, their fracture rate is inversely related to bone mass (Newton-John and Morgan, 1968); but the bone mass of many elderly people who have fractures is normal (for examples *see* Vaughan, 1975) or even above the mean of non-fracturing controls (Exton-Smith, 1976). The changes which lead to osteoporosis and contribute to fracture occur over many years and to a varying extent in different bones. Much of the argument arises from the difficulty of measuring such changes rather than in the interpretation of the results, and it is not possible here to go fully into the many issues which have made age-related osteoporosis difficult to study. This chapter will deal more with the clinical aspects of osteoporosis and describe those forms which are not necessarily related to age in which bone loss is rapid and the cause sometimes unknown.

One should take note of the term osteopenia, which is a definition of reduction in bone mass. It has been used in a number of ways, and is usually applied to that condition which precedes structural collapse or clinical osteoporosis; however, if it is applied to a reduction in mass of mineralized bone it will include osteomalacia (Paterson, 1974) which is confusing. The terms osteopenia and osteoporosis are often used interchangeably.

The reader who requires more information on the various views on osteoporosis should consult the *British Medical Journal* (1971 and 1975) and the *Lancet* (1976), which provides a useful account of some difficulties in osteoporosis research (Table 3.I). He should also consult the useful general article on osteoporosis (Dent and Watson, 1966), which deals with its clinical features, the work of Morgan (1973), *Clinics in Endocrinology and Metabolism* (1973), and the writings of Nordin (1973). Thomson and Frame (1976) provide a recent review of involutional osteopenia, and Gordan (1978) deals with its possible treatment.

TABLE 3.I
Important points about osteoporosis (*Lancet*, 1976)

Long bones gradually increase in external diameter throughout life by periosteal apposition, but endosteal bone is resorbed more rapidly. Therefore cortical thickness decreases with age.

Microradiography in osteoporosis may suggest progressive increase in resorption surfaces with age, but this appearance could be due to a decrease in new bone formation. It is necessary to have details of cells to distinguish active resorption from total resorption.

All age-related changes will correlate with loss of bone mineral.

The symptoms of osteoporosis fluctuate and tend to improve spontaneously. Every advocated treatment appears to improve the disease.

Calcium malabsorption may be the result of osteoporosis rather than its cause.

The role of parathyroid hormone in osteoporosis is undecided.

Recent work strengthens the indications for oestrogen therapy in postmenopausal osteoporosis.

Despite ignorance of the cause of osteoporosis, clear advice on treatment can be given.

CLASSIFICATION

A suggested classification of the types of osteoporosis is given in Table 3.II and discussed subsequently. It can be argued that in all patients with osteoporosis, the causes are multiple and that it is unduly rigid to attempt to classify them into types.

It could also be argued that osteoporosis after the menopause is not idiopathic, and this would reflect current thought. However this would draw a possibly artificial distinction between postmenopausal and elderly osteoporosis, and these

TABLE 3.II
Some causes of osteoporosis

	Unknown − Idiopathic
Common	Postmenopausal
	Elderly
Uncommon	Pregnancy
	Young adult
	Juvenile
	Known or postulated
Immobilization:	local or generalized
Endocrine:	Hypogonadism, including oöphorectomy
	Cushing's syndrome − spontaneous or iatrogenic
	Thyrotoxicosis
	Hypopituitarism
Chromosomal:	Turner's syndrome (XO)
Other:	Rheumatoid arthritis
	Scurvy
	Drugs − heparin, cytotoxic agents
	Inherited − osteogenesis imperfecta
With osteomalacia:	Coeliac disease
	Partial gastrectomy
	Liver disease

two types are therefore considered together (as age-related bone loss). Many classifications of osteoporosis include other causes, more for completeness than for practical reasons (Avioli, 1977). Thus it may be reasonable to include scurvy since there is defective collagen formation consequent upon failure of proline hydroxylation, and osteogenesis imperfecta in which there is an inherited defect of bone matrix formation; but the osteoporosis of rheumatoid arthritis may be merely an example of local immobilization although local bone resorbing factors may be important, and the inclusion of other metabolic bone diseases in which osteoporosis is merely one feature, or of neoplastic disorders of bone, is confusing (Thomson and Frame, 1976).

MEASUREMENT OF BONE MASS

In this chapter it is useful to limit the term osteoporosis to that condition in which reduction of bone mass is associated with structural collapse, and to define the condition which presumably precedes this as osteopenia. Whatever definitions are used, one outstanding clinical problem is how to measure bone loss *in vivo*. If we may make the (sometimes unjustifiable) assumption implicit in the classic definition of osteoporosis that the reduced amount of bone matrix is nevertheless fully mineralized, the measured reduction in bone mineral will indicate reduction in mass. There is in fact no way, except for the direct examination of bone obtained by biopsy, of measuring bone mass clinically other than by its mineral content.

In considering the usefulness of the methods for measuring bone mass (Horsman, 1976; Jowsey, 1977), one must constantly recall that different bones in the skeleton behave differently, and that peripheral bones may lose mass at different rates from central bones. This is particularly notable in the vertebral bodies where spinal osteoporosis is associated with the crush fracture syndrome (Gallagher *et al.*, 1973). Only certain methods can be used to measure bone mass in life. Direct measurement of ash weight per unit volume of bone is not practicable. Bone area on biopsy is subject to very wide variations and samples only a very small part of a heterogeneous tissue. In practice we are left with radiology, photon absorptiometry and neutron activation analysis; in future EMI scanning (p. 59) may also be useful. Single measurements by these methods give a static representation of the amount of mineral present, and sequential measurements may suggest changes in it.

Such sequential measurements may be used in conjunction with more indirect biochemical observations. Thus the hydroxyproline: creatinine ratio and fasting calcium: creatinine ratio in the urine can be used as an indirect measure of bone resorption, and the plasma alkaline phosphatase as a measure of bone formation. Administered isotopes in conjunction with calcium balance can also be used to derive mineralization and resorption rates of bone. Further, although the measurement of bone area is normally very variable, measurement of apparently active cell numbers in the bone may be informative.

In practice much of the difference of opinion lies in the accuracy of x-ray morphometry, densitometry and neutron activation. All of them have disadvantages.

Radiography

Subjective assessment of bone density is very inaccurate. It is only where the density of the bones is almost equal to that of soft tissues that significant loss of mineral can be assumed. The introduction of a standard aluminium step wedge produces some improvement. If the definition of osteoporosis (as distinct from osteopenia) is based on structural collapse, it requires radiological evidence of fracture. Loss of trabeculae may be graded, as in the femoral neck, and this gives a crude index of loss of bone mass (Singh *et al.*, 1972). Jowsey (1977) is of the opinion that to differentiate osteoporosis from normal the Singh index is the most useful measurement.

Measurement of the thickness of the metacarpal cortex from radiographs (morphometry) is a well established way of measuring bone mass, which has been refined by taking the mean of repeated measurements of 3 metacarpals on each hand using accurate calipers and micrometers. It is claimed by some that this method is more accurate than that of densitometry. Further, some contend that the variance of repeated observations is reduced by taking into account some measures of bone size: thus the thickness may be related to the external diameter (D) or length (L) of the bone. The measurement of bone mass and possible indices are discussed in detail by Horsmann (1976) and also by Paterson (1974). In general, unmodified difference between external and internal diameter $(D - d)$ is helpful, although more complex measures such as $D^2 - d^2/DL$ are used (Exton-Smith, 1976).

Densitometry

The absorption by mineral of photons from gamma-emitting isotopes provides an acceptable method of indirectly measuring bone mass. Measurements are often made at the lower end of the radius, but other accessible parts of the peripheral skeleton can be used, such as the metacarpal and calcaneum. The isotope source is either ^{125}I or ^{241}Am. The main source of error in repetitive measurements is difficulty in positioning the arm (or limb) in exact relation to the isotope source or counter. Since calcium will have the same absorbing effect whether it is in mineral or soft tissue, possible errors arise owing to calcification in the soft tissues (as in renal failure). Nevertheless this method is increasingly used with valuable results (Lindsay *et al.*, 1976).

Neutron activation analysis

A small amount of the calcium in the skeleton exists as ^{48}Ca. When this is bombarded with neutrons ^{49}Ca is one of the products with a half-life of 8.8 minutes. Thus rapid counting of the amount of radioactivity due to ^{49}Ca after neutron activation gives an index of the amount of calcium present. Neutron activation may be regional (to the spine or hand, for instance) or to the whole body. Clearly this method relies on complex equipment. So far as the forearm is concerned the difficulties of positioning and of distinguishing between soft tissue calcium and skeletal calcium apply equally to neutron activation and to photon absorption.

TYPES OF OSTEOPOROSIS

The commonest types of osteoporosis are postmenopausal and 'senile', associated with increasing age in females. However because 'senile' has other connotations, 'osteoporosis of the elderly' is preferable). It is convenient to discuss these types under the heading of age-related bone loss and separately from those forms of bone loss which occur at an age where it is not normal to lose bone, i.e. accelerated bone loss. Both types of bone loss are due to an imbalance between resorption and formation, but do not necessarily imply a primary increase in resorption. The particular features of some forms of accelerated bone loss are emphasized later.

Age-related bone loss

Loss of bone with age can be studied in two ways, either by the tedious method of following the changes in a population for many years (a longitudinal study) which provides very useful results on the same individuals as they grow older (Adams, Davies and Sweetman, 1970); or, more usually, by studying the characteristics of different populations at different ages. In the second method it is more easy to obtain information from many patients, but it is never possible

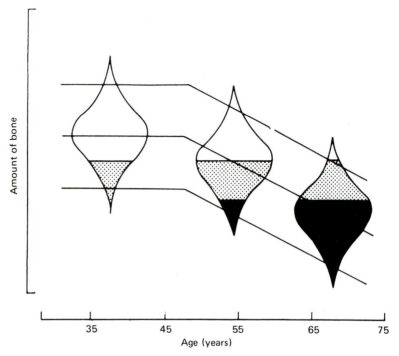

Figure 3.1. To show the way in which a population appears to lose bone with age. The shape shows that values for the amount of bone at different ages have a normal distribution. The stippled area plus the black area indicates −1 SD, and the black area alone − 2.5 SD from the mean at age 30−40 years. (From Newton-John and Morgan, 1970)

to overcome completely the objection that the populations may be different in some important respect. For instance, a group of subjects born five years before another group could have had a different skeleton from the start or have been brought up in a different environment. Although such an objection may be largely theoretical, it cannot be disproved.

Longitudinal studies have shown that in tubular bones formation is periosteal and resorption endosteal (Garn, 1970). In childhood and early adult life bone formation exceeds, or is later equal to, bone resorption. However from the 4th decade, the rate of bone formation appears to decline while resorption continues, and the rate of resorption may increase after the menopause. Thus so far as the long bones are concerned, there is from this age more endosteal resorption than periosteal formation of bone. There is also an increase in intracortical porosity.

TYPES OF OSTEOPOROSIS

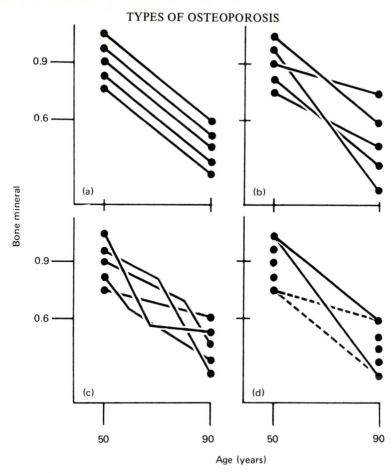

Figure 3.2. To show how individuals may lose bone within a population where the variance about the mean is not increasing with age. Alternatives (c) and (d) allow for differing rates of bone loss in individuals, without an increase in variance; alternatives (a) and (b) do not. (From Smith, Khairi and Johnston, 1975)

The external diameter of the bone slowly increases, but this increase does not keep pace with resorption from inside, and there is therefore a progressive thinning of the cortex. Clearly the process cannot go on at a constant rate since the skeleton would then disappear, and indeed it is known that the individual rate of bone loss, measured by any means, may vary from time to time.

Transverse studies of the bone mass (by any means) of similar populations at different ages show that in women the amount of bone falls more rapidly than that of the men and that this fall appears to begin at the time of the menopause. Subsequent bone loss is progressive, and the mean amount of bone falls but, according to some authors (Newton-John and Morgan, 1970), the variance about this mean remains constant (*Figure 3.1*). Since longitudinal studies on individuals show variable rates of bone loss, these two observations have to be reconciled (Smith, Khairi and Johnston, 1975). It is possible to do so particularly if the rate of loss of bone is proportional to the amount already present; or if the change of rate of bone loss with time is random (*Figure 3.2*).

If it is true that the variance does not in fact change, this implies that there is no separately identifiable sub-population in which the rate of bone loss is excessive. This point cannot yet be considered settled (Exton-Smith, 1976).

It is common knowledge that some bones are different in their structure from others, and thus might be expected to behave differently. This is particularly so when we are comparing the behaviour of a bone which is largely cortical, such as a femur or metacarpal, with one which is largely trabecular, such as the spine. Most measurements of bone mass have been made on predominantly cortical bone and a close relationship to vertebral body collapse is not to be expected.

There are two mysteries in age-related bone loss: first, why does bone formation decrease with age, and second, if indeed bone resorption also increases, especially in women, what is the cause?

There are very few known stimulants of new bone formation, but these include growth hormone (*in vivo*), physical activity and fluoride. It is not known how the stresses of physical activity stimulate bone formation but bone formation rapidly decreases in the immobilized skeleton and immobility of increasing years must play its role in senile osteoporosis. Excessive bone resorption may be an important factor in the loss of bone after the menopause. *In vitro* work (Atkins *et al.*, 1972) showed that the action of PTH on bone was inhibited by oestrogens; and in rats oöphorectomy appeared to increase the sensitivity of bone to parathyroid hormone (Orimo, Fujita and Yoshikawa, 1972), but such work in animals cannot be directly extrapolated to man. The suggestion that osteoporosis in the elderly (as distinct from postmenopausal osteoporosis) results from parathyroid overactivity associated with decline in renal function is difficult to uphold especially since there is little histological evidence of parathyroid overactivity in age-related osteoporosis. Thus we still do not know why bone mass declines with age. Since there is agreement that this bone loss occurs more rapidly and earlier in females than males, one contributing factor is likely to be the rapid decline in female hormones at the menopause (Gordan, 1978). But this cannot be the whole explanation of progressive bone loss in women with age, if only because no hormonal theory explains the wide difference in the racial incidence of this bone disease. Thus the measurements of bone mineral mass in a California community show that it is higher in blacks than whites; and hip fractures are uncommon in black women (*see* Gordan, 1978).

Accelerated bone loss

There are several known causes of accelerated bone loss which require comment. One of the most potent of these is immobilization, either localized or generalized. Studies in astronauts and immobilized medical students have shown the extent of the loss. Medical students put to bed for six weeks and allowed to move in the horizontal plane only (Hulley *et al.*, 1971) showed very rapid bone loss (measured by densitometry) from weight-bearing bones such as the calcaneum. There was hypercalcuria, increased nitrogen excretion and increased hydroxyproline excretion, showing that both bone mineral and matrix were rapidly lost in addition to loss of intracellular muscle protein. When the subjects became mobile again there was densitometric evidence of calcium replacement.

Pace (1977) has reviewed the effects of weightlessness on the skeleton. In astronauts there is continuing loss of calcium during space flight, but Claus-

Walker *et al.*, (1977) found no increase in urine total hydroxyproline. These workers compared the biochemical effects of space flight with those of immobilization in 3 teenage quadriplegic males in whom hydroxyproline and calcium excretion were consistently increased.

Oöphorectomy in young women is associated with accelerated bone loss. Studies of the surprisingly large number of premenopausal women who have had oöphorectomies with hysterectomy show that they lose bone more rapidly than those who have hysterectomies alone (p. 85).

Accelerated osteoporosis also occurs in Cushing's syndrome (iatrogenic or spontaneous) where it is probably due to suppression of synthesis of new bone matrix collagen; in thyrotoxicosis where there is excessive formation and resorption, with disproportionate resorption (breakdown) of collagen; in male hypogonadism where reduction in testosterone leads to defective bone formation; and in chromosomal abnormalities such as Turner's syndrome (XO) where bone formation is defective. In hypopituitarism the causes may be multiple; there are deficiencies of growth hormone, sex hormones and thyroxine.

Occasionally accelerated osteoporosis, which may be severe, occurs in young adults, in women who have recently had an infant, in adolescents or children. The cause in these patients remains quite obscure, and they must be classified as idiopathic.

CLINICAL ASPECTS

Symptoms and signs

The symptoms of osteoporosis are deformity, localized pain and fracture (Chapter 2). The commonest deformity is loss of height owing to vertebral collapse. This is often noted by others rather than the patient, since so few patients know what height they used to be and what their present height is. It is common knowledge that old ladies shrink, and this is due to osteoporosis.

The deformity is often only recognized when clothes no longer fit, owing to apparent descent of the waistline brought on by loss of trunk height. With rapid loss of vertebral height, especially in the young, the head may appear to sink between the shoulders, and the chin comes to rest on the deformed chest, with protrusion of the manubrium sterni.

Most pain in osteoporosis occurs in the back and is associated with vertebral collapse. The pain may come on suddenly with unusual exertion – such as moving heavy furniture. It is severe and localized to the area of vertebral collapse, and the vertebra in this region is often tender to percussion. Fracture over other sites such as the ribs and long bones will naturally also be painful.

In established vertebral osteoporosis with deformity the pain-free periods between vertebral fractures may become less, and are replaced by persistent discomfort (*see* Dent and Watson, 1966, Figure 6).

Examination of the patient with osteoporosis may reveal surprisingly little. It is necessary to have a tape measure and some idea of the normal relationship between span and height (Chapter 2). Progressive loss of trunk height produces a kyphosis and a transverse abdominal crease, with the lower ribs often pressing on the iliac crest.

Physical examination of a patient who has lost height and who is likely to have osteoporosis must include consideration of its cause and exclusion of other disorders. This is particularly so where the 'osteoporosis' is rapidly progressive or in a young person.

Thus excessive bone tenderness and muscle weakness might suggest osteomalacia (p. 104); but in a man in his 50s bone tenderness, anaemia and general ill-health with vertebral fracture is most likely to be due to multiple myeloma (p. 165) or possibly to secondary carcinoma; and, in children, leukaemia may mimic juvenile osteoporosis.

The underlying causes of osteoporosis may often be missed, although they are obvious in retrospect. An example of this is hypogonadism in the male, where the low hairline, smooth skin and highly-pitched voice are clues (*see* p. 91). Similarly, hypopituitarism and Turner's syndrome have their own characteristic features. The patient with coeliac disease (gluten-sensitive enteropathy) may have osteoporosis as well as osteomalacia (p. 120) and have the features of anaemia, tiredness, delayed puberty and sometimes steatorrhoea which go with it. Similarly, osteoporosis may coexist with osteomalacia after partial gastrectomy.

Biochemistry

The values found are, in general, within the normal range. The urine calcium is variable according to the stage of the disease; it may be high during active bone dissolution, especially in the young or during immobilization. More often it is within the normal range or towards its lower limit.

However, indirect measurements of bone resorption, such as the fasting calcium: creatinine and hydroxyproline: creatinine ratio in the urine, may be increased or within the upper end of the normal range, as may be the fasting plasma calcium.

Radiology

The most convincing radiological signs of osteoporosis are provided by structural failure of bone. Measurements of overall density, or loss of trabeculae are more difficult to assess.

In the vertebrae, structural failure causes anterior wedging (*Figure 3.3a*). This affects some vertebrae more than others, and is thus irregular throughout the spine. Since the end plates of the vertebrae tend to collapse (and the intervertebral discs to expand) the vertebrae may become concave. However, except in the young or in inherited osteoporosis (osteogenesis imperfecta), this concavity is neither sufficiently regular nor sufficiently symmetrical for the term 'codfish' spine to be appropriate (*see Figure 3.3b*). Although structural collapse is most marked anteriorly, there may also be posterior vertebral compression.

Before structural collapse occurs various changes are described in bones which become osteoporotic. The vertebrae lose some of the less important trabeculae (the transverse ones) so that the vertical trabeculae become more well defined (Horsmann, 1976) and similar loss of trabeculae occurs in the femoral neck (Singh *et al.*, 1972). In the long bones, metacarpals and phalanges, cortical thickness is reduced.

Bone biopsy

In the elderly patient with osteoporosis in whom there is no clinical or biochemical reason to suspect any other diagnosis bone biopsy is not necessary, but it is important to be aware of the reported high incidence of osteomalacia and to exclude it histologically if necessary. In the young osteoporosis is sufficiently rare that biopsy is necessary to exclude other disorders. Marrow aspiration is essential if leukaemia or myeloma is suspected from the results of routine haematology.

Other investigations

Balance studies and isotope measures of calcium absorption are not necessary for routine diagnosis and treatment of osteoporosis. However it is useful when trying to assess the effect of treatment to distinguish those patients with impaired

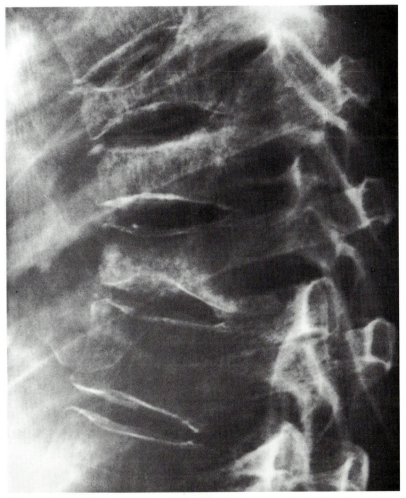

Figure 3.3. (a) The appearance of the spine in osteoporosis. (a) Irregular wedging of vertebrae

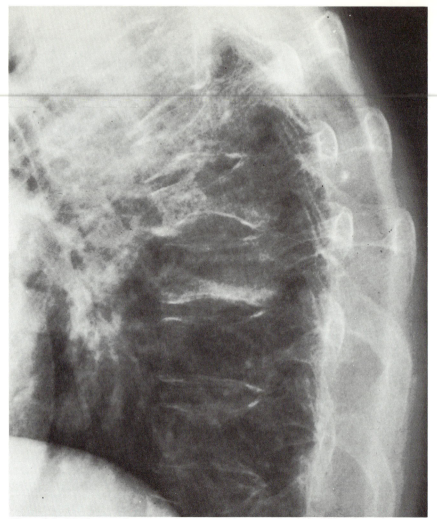

Figure 3.3. (b) The vertebrae may also appear biconcave, but this differs from the appearance in osteomalacia since the deformity is often asymmetric and the vertebrae are not equally affected

absorption of calcium from those without. It is also useful to know the stage of the disease when treatment is begun. This is best achieved by a short metabolic balance study but if this is impossible accurate measurements of 24-hour urine calcium will help. In early active osteoporosis with rapid bone resorption the external calcium balance will be negative and the urine calcium increased. If balance studies and isotope measurements are done simultaneously, figures for rates of resorption and mineralization may be derived (p. 59).

Diagnosis

This has already been alluded to and will be further considered with specific types of osteoporosis.

TREATMENT

Prevention

It is important to prevent as far as possible progressive bone loss with age. Unless the rate of bone loss is proportional to the initial bone mass, which seems unlikely, then there are clear advantages in building up a large skeleton early in life. Although skeletal mass may be in part genetically determined (Smith *et al.*, 1973) the mass of the skeleton also appears to be affected by the amount it is used. Exercise should be encouraged and immobility avoided. If post-menopausal osteoporosis is due to declining female hormone levels, it should be possible to prevent this by giving prophylactic oestrogens, but this problem is not an easy one. Opposing points of view on the administration of oestrogens at this age are given by Studd (1976) and Phillip (1976).

Therapy of established osteoporosis

Certain general lines of treatment are applicable to the common osteoporosis in the elderly. Continued mobility is most important, but this advice is necessarily a compromise since it is usually not possible for a patient who has a painful vertebral fracture to be up and about. Bracing of the spine may provide temporary relief of pain but will also immobilize. In practice the treatment given to an elderly patient with osteoporotic vertebral collapse will often depend on who sees her first; the orthopaedic surgeon will tend to use mechanical support, the physician will aim at mobility, with analgesics for the pain. Accepting this variable approach to initial treatment, other forms of treatment have also been advised, with varying scientific support. These include additional oral calcium, vitamin D and its metabolites, fluoride, anabolic steroids, calcitonin, biologically-active fragments of parathyroid hormone, and intermittent hypercalcaemic infusion.

Oral calcium supplements* are very often given because of the belief that 'senile' osteoporosis is due to calcium deficiency. Certainly it can be demonstrated that in some patients there is poor calcium absorption, and that by giving calcium a temporary positive calcium balance may be produced. Further, if one gives oral calcium late at night this may reduce the fasting calcium excretion in the morning (partly derived from overnight bone resorption). But it is not established that senile osteoporosis is even in part due to calcium deficiency.

Since absorption of calcium in some patients with osteoporosis is poor, some consider it logical to increase it, either by giving vitamin D or its metabolically-active analogue $1\alpha(OH)D$ (*see* p. 122), often in unphysiologically large doses. The need for these larger doses has suggested a partial block in 1α-hydroxylation. To increase intestinal calcium absorption does not in itself represent therapeutic success, especially since the same result would be obtained in normal persons (compare Davies, Mawer and Adams, 1977, and Marshall *et al.*, 1977). It is of great importance to establish where the absorbed calcium goes to, and if it does

---

*Such as 'Sandocal', an effervescent tablet containing 400 mg (10 mmol) of calcium as lactate and gluconate.

go to bone, whether there is an increase in bone mass or an increase in mechanical strength of bone. Since *in vivo* measurements will only record increases in mineral content, they could only give a partial answer, and in any case the effects produced by giving vitamin D or its metabolites to elderly patients with osteoporosis are disputed. Confusion is added when lα(OH)D is given for an apparently mixed bone disease, referred to as 'osteoporosis of ageing' (Lund *et al.*, 1975), in which the clinical and radiological features of osteoporosis are seen but with an increase in osteoid on bone biopsy. At present it is simpler to regard this as osteoporosis with coexistent osteomalacia (which is common in the elderly — Chapter 4), in which a beneficial response to lα(OH)D is to be expected.

Fluoride is one of the few substances which stimulates bone formation and in appropriate doses causes an increase in bone matrix. It has been widely stated that this extra bone matrix is structurally and pathologically unsound and therefore of little use. This is difficult to prove, and the excess osteoid which is a feature of fluoride-treated bones may be mineralized by appropriate doses of vitamin D and calcium. This form of treatment has been particularly supported by the Mayo Clinic group (*see* Jowsey, 1976, 1977); the reader should consult some of the original papers, because the decision to use this form of treatment is not clear-cut.

Anabolic steroids are given because of their supposed effect on protein metabolism. Whilst they may produce a positive nitrogen balance and increase muscle mass, they appear to have little biological effect on collagen, the major skeletal protein of the body.

Recent physiological advances (Chapter 1) suggested that calcitonin might be effective in osteoporosis because of its dramatic effect on bone resorption. However its effect is proportional to the prevailing level of the bone turnover; where this is high, as in Paget's disease, the suppressive effect of calcitonin on bone resorption is considerable; but in persons with osteoporosis the effect is slight, and there is no good evidence that prolonged calcitonin treatment, with the injections and side-effects this entails, prevents loss of bone in elderly osteoporotics.

Parathyroid hormone in very small continuous doses in animals has been shown to have a stimulating effect on bone formation, and data have been presented which suggest that synthetic fragments of the human hormone may decrease bone loss after the menopause (Reeve *et al.*, 1976). This suggestion should be regarded as preliminary although there are some theoretical reasons to support it. Understandably the reader may be confused by the apparently-conflicting views that osteoporosis in the elderly may be partly due to parathyroid overactivity, yet treatment may be effective with synthetic fragments of the human hormone.

A treatment which attempts both to increase calcitonin secretion and to decrease that of PTH is the intravenous administration of intermittent courses of calcium in sufficient doses to produce hypercalcaemia for several hours (Pak *et al.*, 1969). It is very difficult to assess its effects, but it has been claimed to produce a beneficial and prolonged effect on symptoms and on calcium balance. In a subsequent study by Dudl *et al.*, (1973) these beneficial effects were not demonstrated. Popovtzer and his colleagues (1976) combined phosphate supplements with hypercalcaemic infusions, attempting to utilize the effect of phosphate on bone formation. The results illustrated well the pitfalls in the assessment of any treatment for osteoporosis.

Finally, and most importantly, one needs to consider the role of oestrogens, which has been particularly well studied in osteoporosis occurring after an artificial or natural menopause. It is certainly not (yet) appropriate to suggest that osteoporosis in all elderly women should be treated with oestrogens, but the important studies of Lindsay and his colleagues (1976), together with others reviewed by Gordan (1978), strongly suggest that postmenopausal osteoporosis might be prevented by the administration of oestrogens. Lindsay *et al.,* (1976) showed that those women who have oöphorectomies at the same time as their hysterectomies lose bone significantly faster than matched subjects who have hysterectomies alone, and that this loss could be prevented or subsequently reversed by oral oestrogens.

After a naturally-occurring menopause, the situation is less clear. Hormonal measurement together with vaginal histology show that not all women lose oestrogen equally, and this implies that only some women will need or benefit from hormonal therapy. It is probably reasonable to say that if more post-menopausal women received oestrogens, the incidence of 'postmenopausal' osteoporosis would be reduced, but there is a wide divergence about how many such people should be treated and what the likely side-effects will be (Doll *et al.,* 1977). Fowler (1976) particularly emphasizes that since the menopause coincides with a decline in physical activity, osteoporosis occurring at this time could be due to this inactivity. Whilst it still appears to be true that many factors contribute to osteoporosis after the menopause, of which a fall in oestrogen production is only one, the case for prophylactic and therapeutic oestrogens for postmenopausal osteoporosis is becoming stronger (Gordan, 1978).

SOME DIFFERENT TYPES OF OSTEOPOROSIS

Osteoporosis in pregnancy

Young women may occasionally develop the features of osteoporosis during or after pregnancy. Since pregnancy is common, the occurrence of osteoporosis in this group may be coincidental, especially since an equally severe osteoporosis may develop in young men. Dent and Friedman (1965a) supported this view, but Nordin and Roper (1955) suggested that pregnancy and osteoporosis are causally related; this difference of opinion will not easily be settled. It has been pointed out that the calcium 'loss' from the mother attributable to the foetus is only about 20 g but that from continued breast feeding is considerably greater. It may be that the osteoporosis has been present for a long time before the pregnancy, and that the extra mechanical strain to the back of carrying the foetus and, later, the infant precipitates structural collapse of the spine.

In two personally-studied cases (*Figure 3.4*) the rate of loss of height rapidly declined in the first few months after presentation with back pain and loss of stature, which implies that the collapse was self-limiting. Also, calcium balance studies at this time did not show evidence of net loss of body calcium.

Osteoporosis in the young

Apart from that associated with pregnancy, osteoporosis may also be seen in young men and in children during adolescence. The post mortem radiograph in

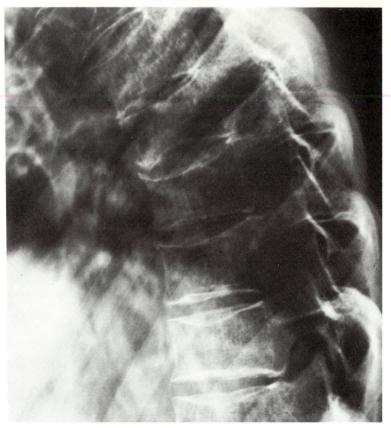

Figure 3.4. Irregular vertebral collapse in a woman aged 26 years, who had lost height and had back pain from the middle of her second pregnancy

Figure 3.5 is that of a man of 29 years, who progressively lost more than 12 inches in height before he succumbed, in whom the cause of osteoporosis was never identified.

Osteoporosis in children and adolescents (juvenile osteoporosis) has certain features which distinguish it from osteoporosis later in life. Amongst the symptoms one of the most prominent is pain at the end of the long bones, especially around the ankles (Lapatsanis, Kavadias and Vretos, 1971). Since onset is most often during the growing phase of early adolescence, failure to grow at the normal rate rather than loss of height may be noticed. Later when the disorder is more advanced, kyphosis and deformity may be obvious. Radiologically, the cortices of the bones are thin and a characteristic appearance is of partial fractures in the metaphyses. Vertebral collapse may be considerable and the shape of the vertebrae may resemble those of osteomalacia, with a regularly biconcave appearance. Severe juvenile osteoporosis, as described by Dent and Friedman (1965b), Dent (1969) and Brenton and Dent (1976) may also show well defined changes from 'normal' to 'abnormal' bone, which is good evidence of the intermittent nature of the disease. In such patients there is no evidence that any specific form of treatment is successful and the disease usually appears to be self-limiting.

Dent and Friedman (1965b) described osteoporosis coming on just before puberty in 6 previously normal children, with fracture of one or more bones as a presenting symptom. Pain around a joint, which might interfere with walking, suggested arthritis in some patients. The radiological changes are very well shown in the original paper. Fractures of the long bones were very common,

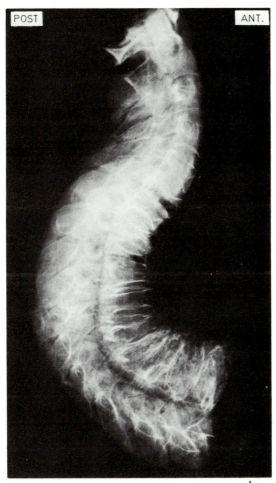

Figure 3.5. Post mortem radiograph to demonstrate the extreme vertebral collapse in a young man who had lost more than 12 inches in height. No cause for the osteo-porosis was found

especially around the hip, knee and ankle joint, beginning as hair-line cracks. The plasma calcium phosphorus and alkaline phosphatase were all recorded as normal for age, but hypercalcuria was seen in some patients. Hypercalcaemia occurred with immobilization. Balance studies showed an inability to absorb dietary calcium. The condition described by Kooh *et al.*, (1973) in 11 children had very similar features, but in 7 of them it appeared to have begun at less than five years of age ('transient childhood osteoporosis of unknown cause'). Again the commonest symptoms were limping, unusual gait or refusal to walk, with severe pain on dorsiflexion of the ankle joint, and radiology showed compression fractures of the distal tibial metaphysis. In 3 patients balance studies showed a negative calcium balance. These authors reasonably point out that since only the

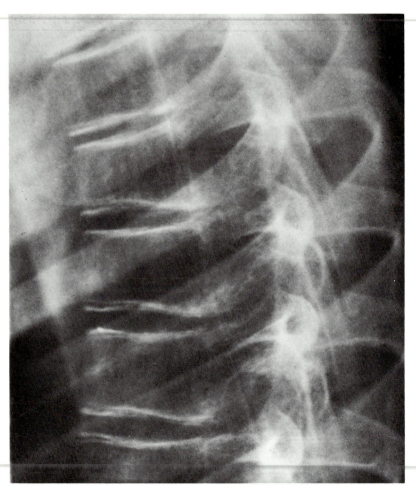

Figure 3.6. The spine in a girl of 10 years with pain in the back and vertebral compression. There was no history of unusual trauma

most severely affected children with this condition are identified, the real incidence may be higher than realized. Bone biopsies in 1 patient demonstrated a diminution in trabecular size and number, with evidence of considerable resorption and arrest of osteoblastic activity. In 7 patients, quantitative microradiography (Jowsey and Johnson, 1972) showed an increase in linear resorption surface, apparently normal formation surface, and a reduction in osteoid width. Some of these patients had values for plasma alkaline phosphatase and urinary total hydroxyproline above that for age-matched controls.

SOME DIFFERENT TYPES OF OSTEOPOROSIS

Severe juvenile osteoporosis is very uncommon and it is more usual to see children with vertebral wedging and pain in the back, which may or may not follow trauma. In such patients the shape of the vertebrae appears to become normal over the following years; calcium balance may be zero or negative at a time when it should be positive. Again it is not clear how often this condition is investigated and how many children who complain of some pain in the back

TABLE 3.III
Clinical details of four adolescents with vertebral collapse

No.	Age and Sex	Symptoms	Ca Balance (mg/day)	Histology	Height
1	10 F	Back pain	+ 10	Not done	+ 1SD
2	10 F	Back pain	+ 245	Normal	Normal
3	12 M	Back pain	− 130	Normal	Normal
4	15 M	Round-shouldered	Not done	Normal	− 1SD

In all patients the spinal radiographs were abnormal.
Patients 1 and 3 do not show the positive calcium balance expected at this age.

are found to have vertebral wedging requiring investigation. The findings in 4 personal cases of this sort are shown in Table 3.III, and a typical radiograph in *Figure 3.6*.

Osteoporosis and endocrine disease

A number of endocrine diseases are associated with a reduction or alteration of bone mass. These include hypopituitarism (and possibly acromegaly), hypogonadism, thyrotoxicosis, and Cushing's syndrome. They are well discussed by Jowsey (1977) and are also considered in Chapter 1.

Hypopituitarism

In hypopituitarism the effects on the skeleton are produced by a number of hormonal deficiencies. Where the pituitary deficiency has been present from childhood growth is retarded and there is delayed or absent sexual development leading to infantilism. The epiphyses remain open, and slow growth may continue for many years. The bones are osteoporotic, and vertebral collapse may occur. Thus an adult with hypopituitarism can give a history of slow continuing growth through early adult life and, later, a loss of trunk height due to osteoporosis.

The physical appearances are characteristic, the patient is dwarfed, has a smooth wrinkled skin, does not shave, has a squeaky voice and a low hairline without frontal recession. The body dimensions are abnormal and trunk height is reduced in proportion to the limbs. It is said that a slipped upper femoral epiphysis is more common in these patients (and those with primary hypogonadism) than normal. This may account for some later appearances of the hip. *Figure 3.7* shows some radiological features. Treatment, which depends on the results of endocrine investigations, includes corticosteroids and thyroxine.

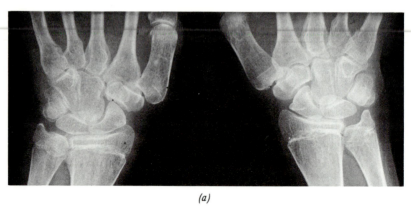

(a)

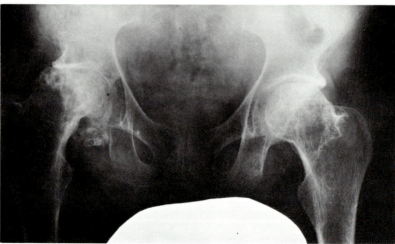

(b)

Figure 3.7. The appearance of the bones in a man of 53 with hypopituitarism. (a) The epiphyseal lines are still present at the wrists. (b) The appearance of the hip joints are unusual, but do not in this case suggest a previously slipped femoral epiphysis

There also seems good reason to give testosterone to prevent progressive osteoporosis and to make the patient feel better, although this is debated.

Hyperpituitarism

Excessive growth hormone will cause gigantism in the growing child and acromegaly if it develops after the epiphyses have fused. Growth hormone itself is ineffective in tissue culture (Chapter 1) and appears to act through the production of somatomedins, substances which primarily stimulate growth of cartilage cells. The patient with acromegaly (*see also* p. 29) may be easily recognizable, although mild cases can be missed. There is enlargement of the soft tissues as well as the skeleton. The facial features become coarse, the nose and tongue large and the hands large and fleshy. The mandible may particularly enlarge, with increased spaces between the teeth. This distortion is shown by x-rays which also show enlargement of the tufts of the terminal phalanges. Most of the skeletal changes can be attributed to increase in subperiosteal bone. Despite this increase, the bones of acromegalics may appear to be osteoporotic. Early extensive arthritis of a degenerative sort is common.

Hypogonadism

In the male with hypogonadism the clinical features are characteristic although the diagnosis is often missed. In contrast to hypopituitarism, such patients are often abnormally tall with relatively long limbs. There is feminine distribution of fat, low hairline and absence of secondary sexual characteristics. Treatment with testosterone* is important and effective.

Turner's syndrome

In this condition, due to the XO chromosome abnormality, osteoporosis is presumably due to lack of sex hormones, with a diminished formation of bone, although it may occur before puberty. Quantitative microradiography also suggests increased bone resorption (Brown, Jowsey and Bradford, 1974), but this method has its limitations. The classic features of Turner's syndrome with growth retardation, webbing of the neck, cubitus valgus, cardiac lesions and lack of sexual development are not always present; chromosomal analysis is therefore necessary to establish the diagnosis. Since many patients are Turner mosaics, with varying proportions of XO and XX cells, it is also necessary to do a full chromosome map from white cells. The examination of a buccal smear for Barr bodies can be misleading.

Thyrotoxicosis

The thyroid hormones have widespread effects on the skeleton (p. 30) and on calcium and collagen metabolism, but these are not often clinically significant. However in thyrotoxicosis excessive bone resorption may lead to osteoporosis, and in some cases to hypercalcaemia and hydroxyprolinuria.

Cushing's syndrome

Cushing's syndrome caused by treatment with corticosteroids is far more common than that which occurs naturally. Its effects on bone are due mainly to suppression of new bone matrix formation, although corticosteroids also affect calcium metabolism. Osteoporosis, often with progressive vertebral collapse, is the commonest skeletal side-effect of such therapy. Fractures of ribs and long bones

*Either as fluoxymesterone orally or testosterone esters by depot injection.

may sometimes heal with excessive callus. Treatment of iatrogenic osteoporosis due to steroids is theoretically easy, by reduction or withdrawal of the drug, but often difficult in practice.

DISCUSSION

The clinical diagnosis of osteoporosis should not be difficult if one restricts the definition to patients with reduced bone mass who have had structural collapse. Difficulties in the measurement of mineral (bone) loss in life have produced much of the controversy in this field, and since the skeleton is heterogeneous in structure, much variation between sites is likely. Some consider that patients with vertebral fracture – the crush fracture syndrome (Gallagher *et al.*, 1973) – have a particular form of osteoporosis. Selective vertebral collapse is likely to be due to the predominance of trabecular bone in the spine compared with cortical bone in the appendicular skeleton, but it does not account for the difference between individuals.

Despite such difficulties, increased use of bone histology and improved methods for bone mineral measurement have brought practical advantages to the patient. Thus the significant occurrence of osteomalacia in elderly osteoporotic patients with femoral neck fractures has been established, and the therapeutic usefulness of oestrogens in preventing or reversing osteoporosis in women who have had bilateral oöphorectomy in addition to hysterectomy has been shown.

It must be our aim to reduce the number of patients who come into the group of 'idiopathic' osteoporosis, but in the majority of elderly patients with osteoporosis it is still not possible to identify the main cause. There are many biochemical alterations in the body with increasing years with which osteoporosis in this age group might be correlated, but it is very difficult to prove that any of them cause osteoporosis.

Assessment of the treatment of osteoporosis is made difficult by its slow progression and the fact that its symptoms are intermittent. Further, the majority of patients are first seen when loss of height and vertebral collapse have occurred, and there is a tendency to improve with or without subsequent treatment. This is often so (although not invariably) even in the more rapidly progressive juvenile osteoporosis.

In terms of numbers the main problem is still the osteoporosis of post-menopausal and elderly women. Since methods now appear to be accurate enough to detect significant loss of bone in this population after 2 to 3 years, it will be possible to make controlled trials of the effect of treatment. One of the main points at issue is whether oestrogens should be given to some or all postmenopausal women and, if so, when this treatment should begin and what are the dangers?

SUMMARY

Reduction of bone mass with age leading to structural collapse is the commonest medical disorder of the skeleton. Since it occurs predominantly in women from the time of the menopause, one important factor is likely to be the decline in oestrogens. Current research is concerned with accurate measurement of this

REFERENCES

bone loss and demonstrating the effectiveness of various forms of treatment. Clinically it remains important to identify the known causes of osteoporosis and to look for others, in an attempt to reduce the size of the 'idiopathic' group.

REFERENCES

Adams, P., Davies, G. T. and Sweetman, P. (1970). Osteoporosis and the effects of ageing on bone mass in elderly men and women. *Q. Jl. Med.*, **39**, 601–615

Albright, F. and Reifenstein, E. C. (1948). *The Parathyroid Glands and Metabolic Bone Disease.* Baltimore: Williams and Wilkins

Atkins, D., Zanelli, J. M., Peacock, M. and Nordin, B. E. C. (1972). The effect of oestrogens on the response of bone to parathyroid hormone *in vitro. J. Endocr.* **54**, 107–117

Avioli, L. V. (1977). Osteoporosis. Pathogenesis and therapy. In *Metabolic Bone Disease.* Eds L. V. Avioli and S. M. Krane. pp. 307–385. London: Academic Press

Brenton, D. P. and Dent, C. E. (1976). Idiopathic juvenile osteoporosis. In *Inborn Errors of Calcium and Bone Metabolism.* Eds. H. Bickel and J. Stern. Lancaster: MTP Press

British Medical Journal (1971). Osteoporosis. (Editorial). **1**, 566–567

British Medical Journal. (1975). Management of osteoporosis. (Editorial). **3**, 307–308

Brown, D. M., Jowsey, J. and Bradford, D. S. (1974). Osteoporosis in ovarian dysgenesis. *J. Pediat.* **84**, 816–820

Claus-Walker, J., Singh, J., Leach, C. S., Hatton, D. V., Hubert, C. W. and di Ferranti, N. (1977). The urinary excretion of collagen degradation products by quadriplegic patients and during weightlessness. *J. Bone Jt Surg.* **59A**, 209–212

Clinics in Endocrinology and Metabolism, (1973). Osteoporosis. **2**

Davies, M., Mawer, E. B. and Adams, P. H. (1977). Vitamin D metabolism and the response to 1.25 Dihydroxycholecalciferol in osteoporosis. *Calcif. Tissue Res.* (suppl) **22**, 74–77

Dent, C. E. (1969). Idiopathic juvenile osteoporosis. *The First Conference on the Clinical Delineation of Birth Defects.* Part IV. Skeletal dysplasias. Birth defects. Original article series, **Vol. V** No. 4. 134–139

Dent, C. E. and Friedman, M. (1965a). Pregnancy and idiopathic osteoporosis. *Q. Jl Med.* **34**, 341–357

Dent, C. E. and Friedman, M. (1965b). Idiopathic juvenile osteoporosis. *Q. Jl Med.* **34**, 177–210

Dent, C. E. and Watson, L. (1966). Osteoporosis. *Post-grad. med. J.* **42**, 581–608

Doll, R., Kinlen, L. J., Skegg, D. C. G., Smith, P. G. and Vessey, M. P. (1977). Hormone replacement therapy and endometrial carcinoma. *Lancet* **1**, 745

Dudl, R. J., Ensinck, J. W., Baylink, D., Chesnut, C. H., Sherrard, D., Nelp, W. B. and Palmieri, G. M. A. (1973). Evaluation of intravenous calcium as a therapy for osteoporosis. *Am. J. Med.* **55**, 631–637

Exton-Smith, A. N. (1976). The management of osteoporosis. *Proc. R. Soc. Med.* **69**, 931–934

Fowler, A. W. (1976). Osteoporosis. *Lancet* **1**, 417

Gallagher, J. C., Aaron, J., Horsman, A., Marshall, D. H., Wilkinson, R. and Nordin, B. E. C. (1973). The crush fracture syndrome in post menopausal women, *Clinics Endocr. Metabolism* **2**, 293–315

Garn, S. M. (1970). *The Earlier Gain and Later Loss of Cortical Bone.* Springfield, Illinois: Charles Thomas

Gordan, G. S. (1978). Drug treatment of the Osteoporoses. *A. Rev. Pharmacol.* **18**, 253–268

Horsman, A. (1976). Bone mass. In *Calcium Phosphate and Magnesium Metabolism.* Ed. B. E. C. Nordin. London and Edinburgh: Churchill Livingstone

Hulley, S. B., Vogel, J. M., Donaldson, C. L., Bayers, J. H., Friedman, R. J. and Rosen, S. N. (1971). The effect of supplemental oral phosphate on the bone mineral changes during prolonged bed rest. *J. clin. Invest.* **50**, 2506–2518

Jowsey, J. (1976). Advances in osteoporosis. *Lancet* **2**, 524–525

Jowsey, J. (1977). *Metabolic Diseases of Bone.* Vol. 1. London: W. B. Saunders Co.

Jowsey, J. and Johnson, K. A. (1972). Juvenile osteoporosis. Bone findings in seven patients. *J. Pediat.* **81**, 511–517

REFERENCES

Kooh, S. W., Cumming, W. A., Fraser, D. and Fornasier, V. L. (1973). Transient childhood osteoporosis of unknown cause. In *Clinical Aspects of Metabolic Bone Disease. Excerpta med.* pp. 329–332

Lancet (1976). Advances in osteoporosis. (Editorial). **1**, 181–182

Lapatsanis, P., Kavadias, A. and Vretos, K. (1971). Juvenile osteoporosis. *Archs Dis. Childh.* **46**, 66–71

Lindsay, R., Aitken, J. M., Anderson, J. B., Hart, D. M., MacDonald, E. B. and Clarke, A. C. (1976). Long-term prevention of post menopausal osteoporosis by oestrogen. Evidence for an increased bone mass after delayed onset of oestrogen treatment. *Lancet* **1**, 1038–1041

Lund, B., Kjaer, I., Friis, T., Hjorth, L., Reimann, I., Andersen, R. B. and Sorensen, O. H. (1975). Treatment of osteoporosis of ageing with 1 α-hydroxycholecalciferol. *Lancet* **2**, 1168–1171

Marshall, D. H., Gallagher, J. C., Guhan, P, Hanes, F., Oldfield, W. and Nordin, B. E. C. (1977). The effect of 1 α-hydroxycholecalciferol and hormone therapy on the calcium balance of post menopausal osteoporosis. *Calcif. Tissue Res.* (suppl.) **22**, 78–84

Morgan, B. (1973). *Osteomalacia, Renal Osteodystrophy and Osteoporosis.* Springfield: Charles Thomas

Muckle, D. S. (1976). Iatrogenic factors in femoral neck fractures. *Injury* **8**, 98–101

Newton-John, H. F. and Morgan, D. B. (1968). Osteoporosis, disease or senescence. *Lancet* **1**, 232–233

Newton-John, H. F. and Morgan, D. B. (1970). The loss of bone with age, osteoporosis and fractures. *Clin. Orthop. Related Res.* **71**, 229–252

Nordin, B. E. C. (1973). *Metabolic Bone and Stone Disease.* London and Edinburgh: Churchill Livingstone

Nordin, B. E. C. (1976). *Calcium Phosphate and Magnesium Metabolism.* London and Edinburgh: Churchill Livingstone

Nordin, B. E. C. and Roper, A. (1955). Post pregnancy osteoporosis. A syndrome? *Lancet* **1**, 431–434

Orimo, H., Fujita, T. and Yoshikawa, M. (1972). Increased sensitivity of bone to parathyroid hormone in ovariectomised rats. *Endocrinology* **90**, 760–763

Pace, N. (1977). Weightlessness. A matter of gravity. *New Engl. J. Med.* **297**, 32–37

Pak, C. Y. C., Zisman, E., Evens, R., Jowsey, J., Delea, C. S. and Bartter, F. C. (1969). The treatment of osteoporosis with calcium infusions. Clinical Studies. *Am. J. Med.* **47**, 7–16

Paterson, C. R. (1974). *Metabolic Disorders of Bone.* Oxford: Blackwell Scientific Publications

Phillip, E. E. (1976). Management of the menopause: two views. Personal view II. *Prescrib. J.* **16**, 58–62

Popovtzer, M. M., Stjernholm, M. and Huffer, W. E. (1976). Effects of alternating phosphorus and calcium infusions on osteoporosis. *Am. J. Med.* **61**, 478–484

Reeve, J., Williams, D., Hesp, R., Hulme, P., Klenerman, L., Zanelli, J. M., Darby, A. J., Tregear, G. W. and Parsons, J. A. (1976). Anabolic effect of low doses of a fragment of human parathyroid hormone on the skeleton in post menopausal osteoporosis. *Lancet* **1**, 1035–1038

Singh, M., Riggs, B. L., Beabout, J. W. and Jowsey, J. (1972). Femoral trabecular pattern index for evaluation of spinal osteoporosis. *Ann. intern. Med.* **77**, 63–67

Smith, D. M., Nance, W. E., Kang, K. W., Christian, J. C., Johnston, C. C. (1973). Genetic factors in determining bone mass. *J. clin. Invest.* **52**, 2800–2808

Smith, D. M., Khairi, M. R. A. and Johnston, C. C. (1975). The loss of bone mineral with ageing and its relationship to risk of fracture. *J. clin. Invest.* **56**, 311–318

Studd, J. (1976). Management of the menopause: two views. Personal view I. *Prescrib. J.* **16**, 51–58

Thomson, D. L. and Frame, B. (1976). Involutional osteopenia: current concepts. *Ann. intern. Med.* **85**, 789–803

Vaughan, J. (1975). *Physiology of Bone.* Oxford: Clarendon Press

4

Osteomalacia

INTRODUCTION

Of all the disorders considered in this book, osteomalacia is probably the best understood in physiological terms. This understanding is particularly due to rapid advances in our knowledge of vitamin D metabolism. Whilst the recognized causes of osteomalacia (Table 4.I) continue to increase from year to year they alter in their frequency according to the efficiency, or otherwise, of preventive medicine. In individuals the correct clinical diagnosis may be long delayed, often because it is not thought of, but the increasing use of bone histology in populations now known to be at risk, such as the elderly, continues to emphasize its frequency. The many recent papers on all aspects of osteomalacia show a laudable enthusiasm concerning this disorder but make it difficult to give a balanced and up-to-date account of its features. In particular, the bone disease associated with chronic renal-glomerular failure has attracted much recent attention. References to specific forms of osteomalacia will be given in subsequent pages but in order to get the subject in perspective the reader should refer to the historical review of Dent (1970), the lecture by Kodicek (1974) and the two-part review of Haussler and McCain (1977).

DEFINITION AND CAUSES

There is a difficulty in the definition of osteomalacia. In this disorder there is defective mineralization of the organic bone matrix, seen histologically in undecalcified bone as an excess of osteoid (*Figure 4.1*). However, excessive osteoid (either as increased thickness of osteoid or increased coverage of bone surface with osteoid) may occur in a number of other conditions such as hypophosphatasia (Chapter 11) or Paget's disease (Chapter 5), and in patients treated with high doses of the diphosphonate EHDP (p. 155). Some workers include these under the heading of osteomalacia. Others (Morgan and Fourman, 1969) restrict the term osteomalacia to that disease caused by vitamin D deficiency. In practice

TABLE 4.I
Causes of osteomalacia and rickets

Vitamin D deficiency ('Nutritional')
        Multifactorial. Poor intake, deficient synthesis

Malabsorption
        Gluten-sensitive enteropathy (coeliac disease)
        Gastric operations
        Bowel resection and bypass
        Biliary cirrhosis

Renal disease
        Renal-glomerular failure
           Renal osteodystrophy
           Dialysis bone disease
     *Renal-tubular disorders
           Familial hypophosphataemic rickets
             (Vitamin-D-resistant rickets)
           Adult-onset hypophosphataemic osteomalacia
           Renal-tubular acidosis
             Inherited and acquired (ureterocolic anastomosis)
           Multiple renal-tubular defects ('Fanconi syndrome')
             Inherited: cystinosis
                         oculocerebrorenal syndrome
                         Wilson's disease
             Acquired: Cadmium poisoning

Anticonvulsant osteomalacia

Tumour rickets

Vitamin-D-dependent rickets

---

*See also Dent (1970) and Dent and Stamp (1977)

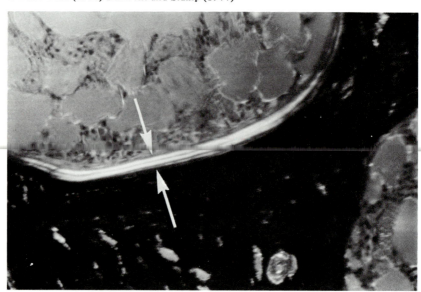

(a)

Figure 4.1. The appearance of normal and osteomalacic bone. (a) In normal bone, the number of birefringent osteoid lamellae is 4 or less and only some bone surfaces are covered with osteoid

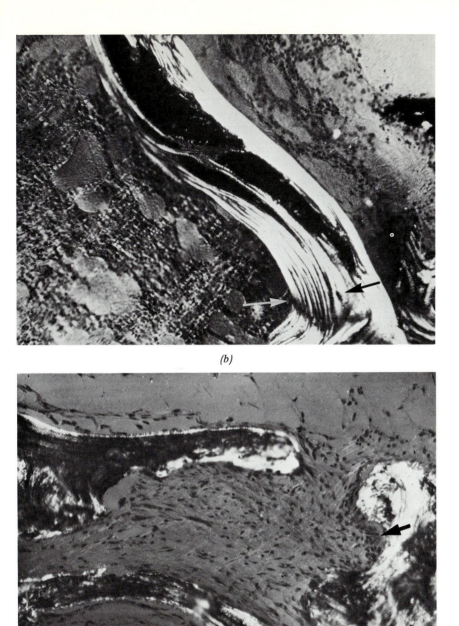

(b)

(c)

Figure 4.1. (b) Bone from a patient with hypophosphataemic osteomalacia. The osteoid seam (arrows) is up to 12 lamellae thick and covers all surfaces. There is no evidence of excessive bone resorption. (c) Bone from a patient with renal-glomerular osteodystrophy. In addition to excess osteoid there is evidence of increased osteoclastic bone resorption (arrow) and fibrosis of the bone marrow. The preparations are undecalcified and seen under polarized light. The osteoid is birefringent. Magnification × 140. Stain von Kossa

it is often most useful (and least confusing) to define osteomalacia (and rickets) as a disorder which results from a lack of vitamin D or an alteration of its metabolism. The clinical features of osteomalacia, its histology, biochemistry and radiology are very similar, whatever the exact cause. Most causes of rickets are the same as those of osteomalacia; rickets refers to the condition when it occurs before growth has ceased, but the terms rickets and osteomalacia are often used interchangeably, irrespective of age.

There is one major exception to this general similarity of cause and clinical features and that is the commonest form of renal-tubular rickets, so called vitamin-D-resistant rickets or hypophosphataemic rickets (Type 1, (Dent)). In this disorder there is virtually no evidence of abnormal vitamin D metabolism and the clinical features are outstandingly different from other forms of rickets. This condition is commonly included in the classifications of rickets, but it can equally well be considered as an inherited low-phosphate disease.

Most of the causes of rickets and osteomalacia are given in Table 4.I, and their main features in Table 4.II. The reader should not be overawed by their number since some are very rare. Clinically, the important major division is into those patients who have osteomalacia owing to vitamin D deficiency, those with malabsorption and those with renal disease. In the last group it is again very important to distinguish between osteomalacia due to renal-tubular disorders and that due to renal-glomerular failure, and not to combine them together under the term 'renal rickets'. Anticonvulsant rickets, tumour rickets and vitamin-D-dependent rickets do not easily fit into these general categories.

VITAMIN D

It is nearly sixty years since the cause of osteomalacia (and rickets) was discovered but we still do not know the precise effects of vitamin D on bone, and exactly how giving vitamin D to patients with rickets cures the bone disease. The one well established fact is that vitamin D aids calcium transport across the small intestine (Chapter 1), but it has many other effects which may not all be limited to transport systems.

The historical development of knowledge about rickets has been well described by Dent (1970) and Dent and Stamp (1977), and clinical and biochemical advances in this disorder have always been closely linked (Kodicek, 1974). Thus the failure of a small proportion of patients with rickets to respond to physiological doses of oral vitamin D or to ultra-violet light led to the discovery and definition of the many causes of 'vitamin-D-resistant' rickets, amongst which were hypophosphataemic rickets and renal-glomerular failure. For many years the cause of vitamin-D resistance in renal failure was unknown, but the striking biochemical advances in the last ten years (referred to in Chapter 1) have provided at least a partial explanation for this and have shown that the metabolically active dihydroxylated derivative of vitamin D, $1,25(OH)_2D$, is a hormone produced by the kidney in response to the requirements of its known target organs.

Alongside such biochemical advances, clinical studies have shown that in the UK the incidence of osteomalacia and rickets due to vitamin D deficiency is increasing, particularly in Asiatic immigrants, and in the elderly. Measurement of plasma 25-hydroxy vitamin D, in which form much vitamin D is stored, has

TABLE 4.II

Main features of different types of rickets and osteomalacia

Type	History	Signs and symptoms	X-ray	Plasma				Urine	Type and daily dose* of vitamin D	Comments
				Ca	P	Pase	Other			
Nutritional	Poor diet Lack of sunshine City dweller	Rickets Osteomalacia	Rickets Osteomalacia	L or N	L	H		Reversible aminoaciduria	From 25 µg (1000 i.u.) vitamin D	Immigrants (Asian) and elderly
Malabsorption	Of coeliac disease, liver disease or gastric surgery	Rickets Features of underlying disorder	Rickets Osteomalacia	L or N	L	H		Reversible aminoaciduria	As for nutritional if coeliac on gluten-free diet	In coeliac disease necessary to give gluten-free diet
Renal-glomerular failure *Renal osteodystrophy*	Of renal disease	Rickets Pigmentation Anaemia	Often gross osteitis fibrosa	L or N	H	H	Uraemia Acidosis	Often proteinuria, inability to concentrate urine	Up to 2.5 mg vitamin D or DHT 1α (OH) D preferable, 1–2 µg	
Dialysis bone disease	On haemodialysis	Myopathy Fractures	Fractures Similarity to juvenile osteoporosis	L or N	H	H	Uraemia Acidosis	On dialysis	Often poor response to vitamin D or metabolites	Cause unknown. Correct induced phosphate deficiency
Renal-tubular disorders *Vitamin-D-resistant rickets*	Family history X-linked	Short stature	Deformity Dense bones Ligamentous calcification	N	L	H (may be N)		Excess† phosphate	0.5–2 mg** vitamin D plus oral phosphate	No myopathy Treat until adult

TABLE 4.II (contd)

Type	History	Signs and symptoms	X-ray	Plasma Ca	P	P'ase L	P'ase H	Other	Urine	Type and daily dose* of vitamin D	Comments
Adult-onset hypophosphataemia	No family history	Myopathy Painful bones	Osteomalacia	N	L		H	May have acidosis	May have excess glycine	Appropriate treatment large dose P ± vitamin D (Dent and Stamp, 1971)	Exclude neoplasm fibrous dysplasia neurofibromatosis
Renal-tubular acidosis	Previous episodes of K deficiency (inherited form) Previous uretosigmoid anastomosis (acquired)	May have dehydration, Na and K depletion	Osteomalacia	N or L	L		H	Hyperchloraemic acidosis	Unable to acidify (distal, classic form)	Large doses NaHCO$_3$ will cure without vitamin D. Vitamin D often given additionally	Nephrocalcinosis and renal stones in inherited form. Uretero-sigmoid anastomosis now little used
Cystinosis	Affected siblings. Well at birth	Severe dehydration and acidosis Photophobia Cystine crystals in the eyes	Rickets	N or L	L		H	May have low Na, K, P, CO$_2$ and increased urea	Aminoaciduria Proteinuria Glycosuria	Vitamin D – up to 1 mg or 1α(OH)D 1 μg, together with phosphate and correction of acidosis	Death in late childhood. Transplantation possible. Cystine crystals in the tissues
Oculocerebro-renal syndrome	Full syndrome in males only	Glaucoma Cataracts Mental deficiency	Rickets	N or L	L		H	Acidosis	Aminoaciduria Organic aciduria	Variable dose vitamin D about 1 mg	Poor prognosis

Disease	History / Clinical features	Bone changes	Other changes	Ca (N/L)	P (L)	Alk. phos. (H)	Urine	Treatment	Notes
Wilson's disease	Family history. May have liver disease or extra-pyramidal signs	Rickets Osteomalacia K-F rings	Rickets Osteomalacia	N	L	H	Amino-aciduria	About 1 mg vitamin D Treat Wilson's disease with penicillamine	Bone disease very uncommon
Cadmium poisoning	Environmental history important	As for osteomalacia		N or L	L	H	Amino-aciduria	Not fully established	Described particularly from Japan. Called ouch-ouch disease (Emmerson, 1970)
Anti-convulsant osteomalacia	Patient with previous or present fits on anticonvulsants	As for osteomalacia Plus underlying disease		N or L	L	H	Amino-aciduria	Up to 1 mg vitamin D	May respond better to 25(OH)D
Tumour rickets	Onset in childhood or adult life. Often no clinical evidence of tumour	Rickets Osteomalacia	Occasional bony tumours	N	L	H	Amino-aciduria	Responds best to removal of tumour, and oral phosphate	Cause unknown. Tumour variable. 'Non-endocrine'
Vitamin-D-dependent rickets	Inherited as recessive	Severe rickets and myopathy	Severe changes	N or L	L	H	Amino-aciduria	Variable. Up to 2 mg vitamin D Rapid response to 1,25(OH)$_2$D, 1 μg	Features of severe nutritional rickets without vitamin D deficiency

* 1 mg = 40 000 iu.
† In relation to plasma phosphate.
** With multiple renal-tubule defects the amount of vitamin D depends considerably on the correction of the acidosis and hypophosphataemia. Note that changes in plasma Ca (low (L) or normal (N)), P (low (L)) and alkaline phosphatase (high (H)) are the same for most causes of rickets. Likewise the urine calcium is often very low (less than 1 mmol (40 mg) a day) but may be normal in the renal-tubular disorders and high in iatrogenic hypophosphataemia

shown how low the values may fall particularly in late winter; and it has demonstrated and re-emphasized the importance of cutaneous synthesis of vitamin D from its precursors under the influence of sunlight (Preece *et al.,* 1975; Stamp, 1975; Stamp, Haddad and Twigg, 1977). Despite the explosive rate of research, many apparently simple problems remain unanswered; for instance, if 1,25 $(OH)_2D$ is the main active metabolite of vitamin D produced by the kidney, why do not all binephrectomized patients (on chronic haemodialysis) develop osteomalacia? (*See* page 123); and if the main demonstrable action of 1,25 $(OH)_2D$ on bone is to increase resorption, how does vitamin D heal rickets?

The salient points of vitamin D metabolism have been described (Chapter 1), and the biochemical effects on its target organs are well demonstrated when patients with rickets are treated with vitamin D (p. 114).

The possibility that vitamin D has an effect on muscle is suggested by the proximal myopathy of osteomalacia (Schott and Wills, 1976). When muscle contracts the thin (actin) and thick (myosin) filaments slide past each other, and this interaction requires a certain concentration of calcium. The supply of calcium is controlled by the sarcoplasmic reticulum, which concentrates calcium within itself when relaxation occurs (hence the name 'relaxing factor'), and releases it during contraction. In muscles from vitamin-D-deficient animals this concentration appears to be defective (Curry *et al.,* 1974). Some evidence (Birge and Haddad, 1975) suggests that 25(OH)D has a direct action on the protein metabolism of muscle. Muscle, in common with other tissues, contains a 25(OH)D binding protein but it is difficult to know the significance of this.

The application of the new knowledge about vitamin D to the known clinical causes of osteomalacia has been very important but not invariably illuminating. The accompanying diagram (*Figure 4.2*) superimposes these causes on the metabolic paths of vitamin D. Thus, the intake of vitamin D in food may be deficient in the elderly and the immigrant (1); the absorption may be defective in coeliac disease or biliary obstruction (2); phytate and chapattis are said to affect calcium (but not vitamin D) absorption (3); exposure of the precursors in the skin to sunlight is reduced in northern cities, in the elderly and immigrants (4); pigment has been thought by some to be important in reducing cutaneous vitamin D production (5); anticonvulsants may interfere with normal hepatic 25-hydroxylation (6) and finally $l\alpha$-hydroxylation is reduced or prevented by renal failure (7), nephrectomy (8) hyperphosphataemia (9) and hypoparathyroidism (10), and is probably deficient in vitamin-D-dependent rickets (11).

It is important to recall that in many patients osteomalacia is multifactorial; for instance in the elderly person, poor vitamin D intake, poor exposure to sunlight, anticonvulsant treatment and increasing renal failure may coexist.

In renal failure, advances in vitamin D metabolism have had particular relevance in explaining both the known impairment of intestinal calcium absorption and the occurrence of a form of bone disease with a combination of defective mineralization and excessive resorption (*Figure 4.3*). When glomerular filtration rate falls, two separate events occur; the plasma phosphate tends to increase, there is consequent fall in plasma calcium, PTH secretion increases and so does bone resorption; at the same time, $1,25(OH)_2D$ production falls, leading to impaired intestinal transport of calcium and defective mineralization of bone. The combination of defective mineralization and excessive resorption leads to the particular features of renal-glomerular osteodystrophy (p. 121).

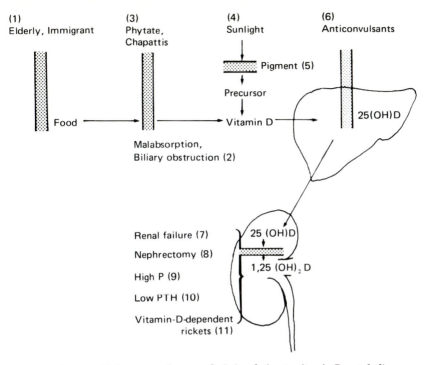

Figure 4.2. The likely causes of osteomalacia in relation to vitamin D metabolism. For further explanation see text; and for nomenclature of vitamin D metabolites see Chapter 1. For clarity the enterohepatic circulation of 25(OH)D is omitted

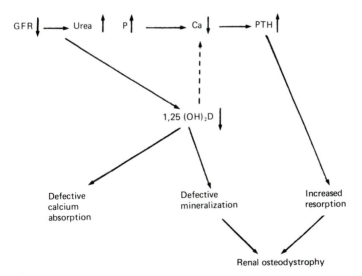

Figure 4.3. Proposed mechanisms for the production of osteo-dystrophy in renal-glomerular failure, Additional arrows could be drawn. Thus the increase in phosphate may reduce 1α-hydroxylation and the increase in PTH would increase it

It is not possible, or necessary, to refer to all the recent work on vitamin D metabolism. However the reader should always bear in mind the importance of the cutaneous synthesis of vitamin D. Thus comparison of the circulating levels of 25(OH)D after ultraviolet irradiation and oral vitamin D suggests that the endogenous production rate of vitamin D is equal to an oral daily dose of 8000—10 000 iu. (Stamp, Haddad and Twigg, 1977).

CLINICAL FEATURES

Because of their overall similarity the clinical features of osteomalacia can be dealt with together. Individual differences will be considered later in this chapter.

Symptoms

The main symptoms of osteomalacia are bone pain and tenderness, deformity and proximal muscle weakness. They are superimposed on the features of the underlying disorder and may be accompanied by symptoms of hypocalcaemia. In the elderly, these features may not be obvious and diligent enquiry is necessary.

Characteristically, all bones are painful and tender; the tenderness may be sufficient to wake a patient turning over in bed at night; often tenderness is most marked in the lower ribs; and sometimes it is increased over a Looser's zone.

Deformity is most marked when the disorder comes on during childhood.* The classic deformities of rickets are well described (see Nelson, Vaughan and McKay, 1969), with bowing of the long bones, a rickety rosary, Harrison's sulcus and bossing of the frontal and parietal bones. Progressive deformity often with knock knee may also be very marked in the rapid growth of adolescence, particularly when bone resorption is excessive, as in renal osteodystrophy (*Figure 4.4*). Osteomalacia may produce a triradiate pelvis (*Figure 4.5*) and later in life a gross kyphosis and corresponding deformities of the chest.

Weakness of the proximal muscles is an important symptom (Smith and Stern, 1967), but its extent may vary from one type of osteomalacia to another, and it may be difficult to decide how much the symptoms are due to myopathy or to bone pain. The commonest feature is a waddling gait; this may be associated with difficulty in getting up (and down) stairs, out of low chairs and in and out of small cars. The weakness is most commonly noticed in the legs, but is also present in the arms; trunk weakness may make it difficult to get out of bed, and the child with rickets goes off his feet. Whilst the weakness may be profound enough to suggest muscular dystrophy, sometimes the only complaint is that the muscles are stiff, or that it is difficult to get about. In the elderly, paraplegia may be falsely diagnosed (Smith, 1976).

There may be clues to the underlying disorder: for instance, the anaemia, tiredness or diarrhoea (in the minority) of coeliac disease; the nocturia and thirst of chronic renal failure; or the history of epilepsy in a patient with anti-convulsant osteomalacia.

Occasionally, tingling of the hands and feet, or tetany due to hypocalcaemia, may be the presenting symptom. If so, its significance may be unrecognized and it is likely to lead the patient to a psychiatrist.

*Dent and Stamp (1977) explain the distribution of the deformities of rickets at different ages.

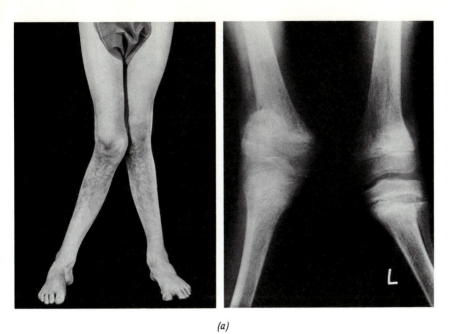

(a)

(b)

Figure 4.4. Severe knock knees (a) and obvious rickety rosary due to enlarged costochondral junctions (b) in a boy with renal osteodystrophy from renal failure due to urethral valve obstruction. Photographs and corresponding x-rays are shown together

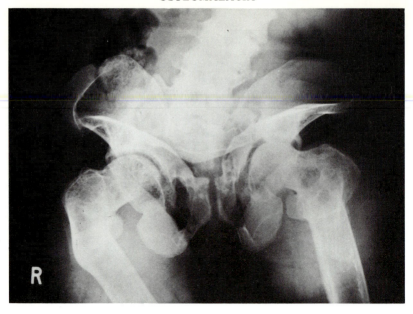

Figure 4.5. Severe triradiate deformity of the pelvis in a patient with vitamin-D-dependent rickets

Signs

Examination of the patient with rickets or osteomalacia will confirm the symptoms. The body proportions should be measured (Chapter 2). Patients with lifelong bone disease, such as Type 1 resistant rickets, may have relatively short legs. Late-onset osteomalacia will produce a relatively short trunk. Latent tetany should be looked for. Physical examination may yield important clues about the underlying disorder (Chapter 2). Examples are abnormalities in the eyes, the scars of previous gastric surgery, or the combined pallor and pigmentation of renal failure.

The presumptive clinical diagnosis of osteomalacia or rickets is confirmed by biochemistry, radiography, histology if necessary, and by response to treatment.

Biochemistry

The biochemistry of patients with osteomalacia varies with the cause (Table 4.II). This is important in diagnosis, and should prevent clinical mistakes based on generalizations, for instance the exclusion of osteomalacia because the plasma calcium is normal (Smith, Lindenbaum and Walton, 1976; *see also Figure 4.8*). The commonest changes, as in patients with osteomalacia due to vitamin D deficiency or to malabsorption, are described.

Plasma

The total calcium corrected for protein may be low or normal; the inorganic phosphate is nearly always low and the alkaline phosphatase is usually increased.

The biochemical changes depend partly on the stage of the disease, which can be artificially divided into three stages of severity (Albright and Reifenstein, 1948), and partly on the degree of parathyroid overactivity. For instance, some patients with gluten-sensitive enteropathy may have evidence of considerable secondary hyperparathyroidism, with normocalcaemia, hypophosphataemia and increased alkaline phosphatase whereas in others the plasma calcium remains low, the phosphate is high and there is little evidence of bone resorption. The biochemistry also depends on age (Chapter 2). There are now established normal ranges during growth and early adult life but the interpretation of biochemical findings in the elderly can be difficult. Thus there are many causes of a raised alkaline phosphatase at this age, to which osteomalacia contributes a small fraction, and in elderly patients with osteomalacia the alkaline phosphatase can be normal. A raised blood urea, a low plasma calcium and, in females, an increase in plasma phosphate (Hodkinson *et al.*, 1973) also occur with age.

Amongst students there is widely held belief that in osteomalacia the plasma phosphate should be high because the plasma calcium is low. This is because the constancy of the Ca × P product has been overemphasized by physiologists. In fact there are many situations where this reciprocal relationship between Ca and P does not occur. The reduction in plasma P is probably the most useful and consistent indicator of osteomalacia, except in patients with renal-glomerular failure. This hypophosphataemia reflects the combined effects of vitamin D deficiency and parathyroid overactivity.

The patient with vitamin D deficiency may show a reversible hyperchloraemic acidosis.

Urine

The urine calcium is low, and can be reduced to less than 10 mg (0.25 mmol) daily. The normal lower limit of urine calcium must be related to the prevailing level of plasma calcium (Chapter 2). The urine phosphate is also increased in relation to that in the plasma, owing to the reduced renal-tubular reabsorption of phosphate associated with parathyroid overactivity. Vitamin D deficiency also produces a reversible amino-aciduria and an inability to acidify the urine.

In the less common forms of osteomalacia, other biochemical changes may occur (Table 4.II).

Radiology

In rickets, the main abnormalities are at the growing ends of the long bones, where the organization of the growth plate and the normal processes of mineralization are disorderly. Radiologically, the growth plate appears widened; and the metaphysis is widened, cupped and ragged (*Figure 4.6*). In renal osteodystrophy bone resorption is excessive. The appearance has been referred to as that of a 'rotting stump' (*Figure 4.7*).

In osteomalacia, there may also be deformity with, for instance, a triradiate pelvis (*Figure 4.5*), but the radiological hallmark of active osteomalacia is the Looser's zone (*Figure 4.8*) which is a ribbon-like area of demineralization. This may occur in any bone but most often in the long bones, the ribs and the

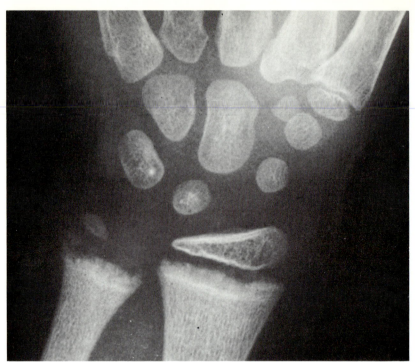

Figure 4.6. The radiological appearance of rickets in the wrist of a child with cystinosis. The growth plate is widened and the metaphysis is cupped and ragged

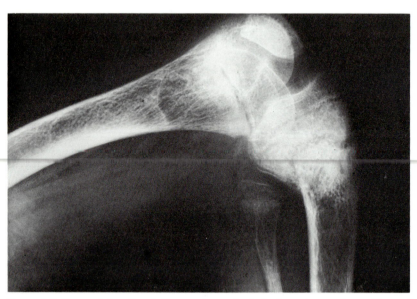

Figure 4.7. Severe deformity at the growing end of the bone in a patient with renal osteodystrophy, owing to a combination of osteomalacia and osteitis fibrosa. This appearance may be referred to as a 'rotting stump'. The tibial metaphysis has fractured. There is also calcification of the large and small blood vessels, visible on the original x-ray

pelvis. In contrast to the multiple microfractures of Paget's disease, Looser's zones occur on the concavity of long bones such as the femur where they are wider and single. Although it is unusual to have multiple Looser's zones in one bone, several bones may be affected and the Looser's zones may be bilateral and symmetrical. Dent and Stamp (1977) suggest they follow trauma.

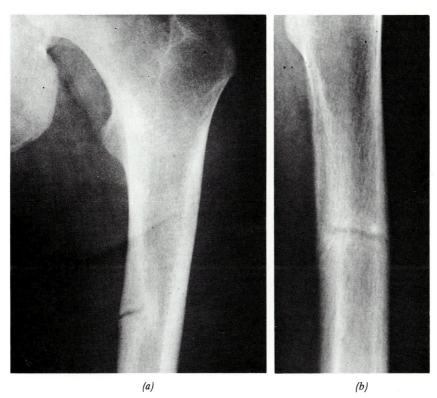

(a)   *(b)*

Figure 4.8. Characteristic appearance of a Looser's zone in the left femur of a patient with hypophosphataemic osteomalacia: (a) anteroposterior; (b) lateral view. The plasma calcium was normal, and phosphate 1.7 mg per 100 ml

Other changes in osteomalacia include deformities of the chest and spine. The vertebral bodies are uniformly and symmetrically affected and may be biconcave *(Figure 4.9a)*; when present this regular biconcavity suggests osteomalacia rather than osteoporosis *(see* Chapter 2).

The radiological appearances of the bones may differ according to the cause of the rickets (or osteomalacia). In all except those due to inherited or acquired hypophosphataemia, where the plasma calcium is normal, there is a variable degree of bone resorption attributable to parathyroid overactivity. Resorption is most marked in the young and in renal osteodystrophy, and occurs in a number of places *(see* Chapter 6), which include the symphysis pubis, sacroiliac joints, femoral neck and medial border of the proximal end of the tibia, but it is often best seen in the proximal phalanges (radial borders rather than ulnar) and the tufts of the distal phalanges.

109

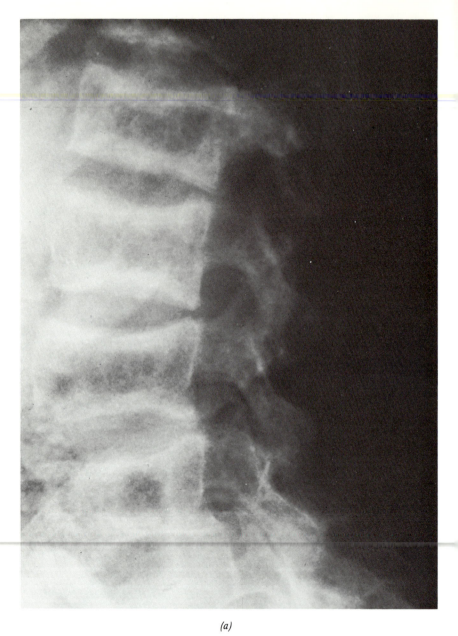

(a)

Figure 4.9. The appearance of an osteomalacic spine. (a) The vertebrae are
symmetrically biconcave and appear to be equally affected

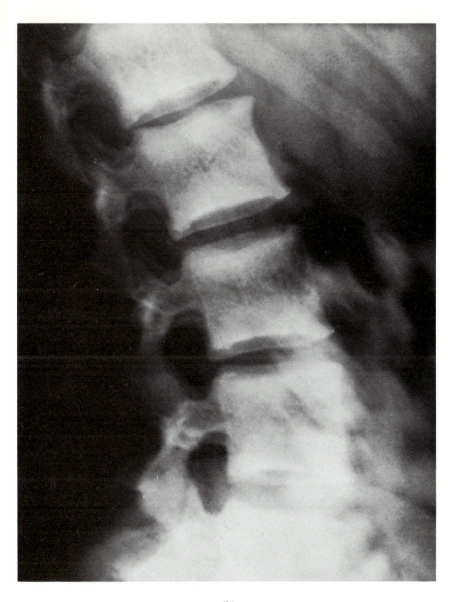

(b)

Figure 4.9. (b) Increased density of the end plates may produce the so-called rugger jersey sign, supposedly only in renal-glomerular osteodystrophy

111

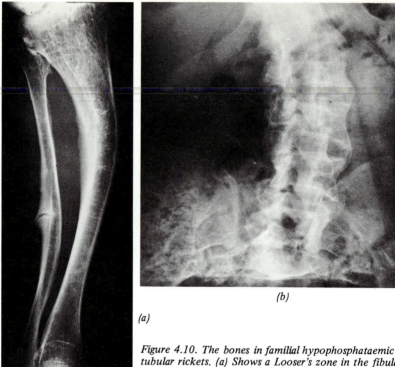

(a)

(b)

Figure 4.10. The bones in familial hypophosphataemic renal-tubular rickets. (a) Shows a Looser's zone in the fibula with buttressing and deformity of the tibia; (b) demonstrates ligamentous calcification in the spine, which in this patient led to paraplegia

Excessive bone formation and osteosclerosis together with resorption produce the dense end plates of renal osteodystrophy (rugger jersey spine) (*Figure 4.9b*). *Figure 4.9a* demonstrates this starting to occur in a patient who had no significant renal failure.

The radiographs of X-linked hypophosphataemic rickets differ from other forms of rickets. In addition to the absence of resorption they may be abnormally thick and the long bones are short and buttressed (*Figure 4.10a*). Exostoses and calcification at muscle insertions occur. Rarely the spinal ligaments become mineralized, with an appearance similar to ankylosing spondylitis (*Figure 4.10b*).

Histology

Histological confirmation of rickets or osteomalacia is not always necessary. Thus in the child with 'nutritional' rickets or coeliac disease the diagnosis may be clinically obvious, and readily confirmed by biochemistry and radiology. Where the diagnosis is in doubt and where, for instance, it is particularly important to exclude osteomalacia in a patient who is known also to have osteoporosis, bone histology may be very useful. It is also useful as a guide to treatment. In contrast to osteoporosis or Paget's disease, osteomalacia affects the whole of the skeleton. Despite the differences of opinion amongst bone pathologists about the minimum requirements for the diagnosis of osteomalacia, in the individual case the diagnosis is not usually difficult. Undecalcified cancellous bone from such a patient will show both that the osteoid is increased in thickness and that it covers more of the mineralized surfaces than normal (*Figure 4.1*). There will be variable bone resorption and increase in fibrous tissue. In those conditions where hypophosphataemia is the main abnormality, osteoid is excessive but resorption is not increased.

Apart from osteoid, osteomalacic bone should show a reduction or absence of the calcification front (Bordier and Tun Chot, 1972), but this may be difficult to demonstrate. Similarly, bone that is not mineralizing will fail to show the normal uptake of tetracycline.

Other investigations

The foregoing investigations are all that are required for the diagnosis of osteomalacia, but not all are necessary in every case. However some additional measurements may be needed to diagnose the exact cause.

For example (Table 4.II) the finding of severe acidosis in a child might suggest cystinosis; a raised blood urea points to renal-glomerular bone disease; a generalized amino-aciduria (which persists after treatment) suggests multiple renal-tubular defects (the Fanconi syndrome); and a low caeruloplasmin level will support the diagnosis of Wilson's disease.

Since measurements of vitamin D metabolites are increasingly available, they have been used to elucidate the biochemical causes of osteomalacia (Haussler and McCain, 1977). The easiest of these measurements is the circulating 25(OH)D. In general this reflects the vitamin D content of the body and it is lower than normal in patients with 'nutritional' osteomalacia. It is however important to note that the normal range varies with the season, being highest in the summer and autumn, and that 25 (OH)D may be undetectable in the plasma in the winter, without clinical osteomalacia. It is also low in liver disease (p. 120) and in some patients with renal failure.

The excretion of total hydroxyproline (THP) in the urine is often increased in osteomalacia, in approximate proportion to the degree of secondary hyper-parathyroidism. There is normally a considerable peak of THP excretion in adolescence (Chapter 2). Measurement of THP excretion does not help to diagnose osteomalacia although it may help in assessing treatment.

In routine diagnosis and treatment a calcium balance is superfluous; however it may give information on the effectiveness of treatment, such as $1,25(OH)_2D$ in renal failure, which is not otherwise available. A method which has been used in the past to support the diagnosis of osteomalacia is an intravenous calcium load more of which is retained by the osteomalacic patient than the normal subject. Radioactive isotopes of calcium have been used to study calcium absorption and metabolism but controversies exist about the interpretation of the results (Chapter 2). The total absorption of an oral dose may be measured with a whole body counter; as an example this method has been used to demonstrate the effectiveness of microgram doses of $1,25(OH)_2D$ in increasing calcium absorption in renal osteodystrophy (Henderson *et al.*, 1974).

Response to treatment

The rapid response of a patient with vitamin D deficiency rickets or osteomalacia may confirm the diagnosis. Clinically, the most rapid response may be an increase in muscle strength within a few days; in contrast, skeletal pain may temporarily

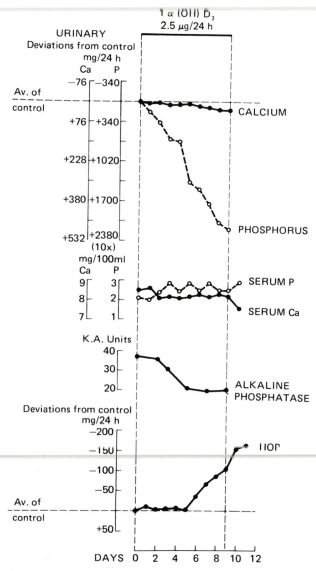

Figure 4.11. The biochemical changes which follow treatment of osteomalacia with crystalline 1α(OH)D. The deviations from baseline are plotted cumulatively and increased excretions are shown as negative values. The main changes are a decrease in urine phosphate and an increase in total hydroxyproline (HOP). (Reprinted by permission from Bordier et al., (1974), New England Journal of Medicine 291, 866–871)

worsen, associated with the biochemical evidence of increased bone cell activity. Biochemically there may be a temporary increase in plasma alkaline phosphatase (the phosphatase 'flare'); total urine hydroxyproline temporarily increases; PTH level falls and 25(OH)D levels rise (Preece *et al.*, 1975). With the fall in PTH there is an increase in plasma P, presumably due to an increase in TmP (p. 51). The effect of crystalline lα(OH)D has been analysed in detail by Bordier *et al.*, (1974); *Figure 4.11* is reproduced from this paper. The reader should note that the results are plotted cumulatively, which exaggerates them. Improvement in radiological appearances may be seen within a month of effective treatment, but the histology may be abnormal for months. The treatment of osteomalacia depends on its cause and may be excessively simple or very difficult. It is useful to remember that 1 mg of vitamin D is equivalent to 40 000 iu. A detailed and useful description up to 1972 is given by Parfitt and Frame (1972), and recently by Dent and Stamp (1977).

In vitamin D deficiency a response may be obtained by a daily dose of a few hundred units daily, but it is often useful to give more (for instance, 1,25 mg daily for a week). Lack of response to doses of vitamin D in the range of 10 to 100 μg (400—4000 iu.) suggests that the osteomalacia is not due to simple deficiency of vitamin D and, providing the histological diagnosis is sound, suggests some underlying cause such as coeliac disease or renal failure, or some increase in vitamin D requirement, such as anticonvulsants. The treatment of these conditions varies, and will be dealt with subsequently.

DIFFERENTIAL DIAGNOSIS

The diagnosis of osteomalacia presents few problems once it is thought of, but there is often a considerable delay since it is mistaken for many other conditions. Thus an Indian immigrant who complains of stiffness of the muscles and difficulty in walking is far more likely to be sent to the department of rheumatology or physical medicine than to be investigated for osteomalacia; the young patient with coeliac disease whose main symptom is proximal muscle weakness is likely to be investigated first by a neurologist; and spontaneous tetany, because of its rarity, is often unrecognized. Since pain in the bones is sometimes mistakenly located in the joints or soft tissues, non-specific diagnoses of 'rheumatism' or 'fibrositis' are common.

Osteomalacia should be differentiated from other forms of metabolic bone disease, from other causes of proximal muscle weakness, from other causes of painful bones, and from psychological illness. The characteristic biochemical changes and the information in the appropriate chapters should distinguish it from other metabolic bone diseases; nevertheless osteomalacia may coexist with osteoporosis, Paget's disease or primary hyperparathyroidism, and bone biopsy may be necessary to elucidate the problem. So far as other causes of proximal muscle weakness are concerned, polymyalgia rheumatica, dermatomyositis and polymyositis, neoplastic neuromyopathy and unusual forms of muscular dystrophy may need to be distinguished. Confusion can again be avoided if all patients with this symptom have measurements of plasma calcium, phosphorus and alkaline phosphatase, recalling that the rare patient with adult-presenting hypophosphataemia may have the most severe weakness.

OSTEOMALACIA

The distinguishing features of polymyalgia rheumatica, which is a relatively common disorder, are onset in late middle age, general ill-health, painful and weak proximal muscles with inability to raise the arms above the head, a considerably increased ESR and rapid response to corticosteroids. The child with dermatomyositis or X-linked muscular dystrophy has none of the features of rickets apart from the myopathy. In adults the varying neurological syndromes associated with neoplasm may provide temporary difficulty.

The tenderness of bones in osteomalacia may be simulated by that of leukaemia, which can rapidly be distinguished by the peripheral blood count and bone marrow examination, and of myeloma, in which case radiography and examination of the urine for light chains will also be necessary.

Finally, a number of the symptoms suggesting osteomalacia may occur in psychological illness, namely pain, weakness and abnormal gait. Such patients will not have abnormal biochemical findings.

DIFFERENT TYPES OF OSTEOMALACIA

The following pages will deal in more detail with the causes and clinical aspects of different types of osteomalacia. Osteomalacia may also occur in association with other common conditions such as rheumatoid arthritis (Maddison and Bacon, 1974), but these are not considered separately. Dent and Stamp (1977) provide a comprehensive account.

Nutritional osteomalacia

Nutritional osteomalacia (or rickets) is usually due to a combination of poor dietary intake of vitamin D and defective cutaneous synthesis of vitamin D following reduced exposure to ultraviolet light. The relative contributions from these sources is still the subject of discussion, but recent work has emphasized the considerable importance of endogenously synthesized vitamin D.

There seems little doubt that the frequency of nutritional osteomalacia is increasing, but the extent of this increase is difficult to establish and there are wide differences between regions and investigators. Thus, for instance, if we use excess osteoid coverage, rather than thickness, to diagnose histological osteomalacia, up to 30 per cent of elderly patients with fractures of the femoral neck appear to have this disorder (Aaron et al., 1974); and the extent of osteomalacia amongst adolescents may appear to be excessively high when based on levels of plasma alkaline phosphatase (Cooke et al., 1973) if due account is not taken of the increased values normal for this age (Round, 1973).

In the UK nutritional osteomalacia occurs particularly amongst the elderly, and in Asian immigrants at all ages.

Osteomalacia in the elderly

Osteomalacia in the elderly is a disabling disease. Affected patients may find it very difficult to get about, and at this age osteomalacia is a significant cause of undiagnosed inability to walk. This disability is probably in part due to proximal muscle weakness, and is not necessarily related to the presence of pseudofractures. In the elderly the high incidence of osteomalacia may be due to many causes — particularly reduced exposure to sunlight ('osteomalacia of the housebound', Hodkinson, *et al.*, 1973), and poor dietary intake, but also to often-prescribed anticonvulsants. The incidence of histological osteomalacia in patients with femoral neck fracture is said to be highest in the late winter and spring (Aaron, Gallagher and Nordin, 1974), when 25(OH)D levels are lowest (McLaughlin *et al.*, 1974).

Chalmers (1970) drew attention to the frequency of subtrochanteric fractures of the femoral neck in patients with osteomalacia and Jenkins *et al.*, (1973) noted the high incidence of biochemical and histological osteomalacia in patients over the age of 70 years but found no close coincidence between the histologically and biochemically defined groups and no significant difference in incidence between those with trochanteric and those with subcapital fractures. In contrast O'Driscoll (1973) found that a particular type of subcapital fracture was associated with osteomalacia.

The Leeds group (Aaron *et al.*, 1974) have recorded a high incidence of osteomalacia in femoral neck fractures but made the additional suggestion that at least some of the coexistent osteoporosis might be due a degree of vitamin D deficiency. In Denmark the plasma 25(OH)D levels in 67 consecutive cases of femoral neck fracture were found not to be significantly different from control patients at the same time of the year (Lund, Sorensen and Christensen, 1975), and the intake of vitamin D is much higher in the elderly population than in England. These authors suggest that the recorded high incidence of histological osteomalacia in patients with osteoporosis was due to a defect in the renal conversion of 25(OH)D to 1,25(OH)$_2$D associated with declining renal function with age. The term 'osteoporosis of ageing' has been used where there is osteoporosis with histological evidence of osteomalacia (Chapter 3). This concept is confusing, and the original papers should be studied.

The practical message is that it is always important to exclude osteomalacia in an elderly person with fracture; since there are many such elderly patients and full investigation is not always necessary or desirable in doubtful cases, the answer may be most easily obtained by a therapeutic trial of vitamin D. It has been suggested that such patients have an increased requirement for vitamin D but there seems no clear reason to give vitamin D metabolites or analogues; a short course (2–3 weeks) of oral calciferol in a dose of 1.25 mg daily should produce measurable biochemical and clinical improvement in a patient with osteomalacia (and will improve vitamin D stores). Vitamin D should not be continued at this dose (or replaced with even larger doses by injection for the sake of convenience) but should be reduced to physiological levels, such as vitamins A and D capsules, BPC, two daily, providing 900 iu. or about 23 µg.

Osteomalacia and rickets in the immigrant

There appear to be good reasons why Asian immigrants should develop nutritional osteomalacia, since they tend to live in Northern cities well away from sunlight and continue to take a diet deficient in vitamin D. The incidence in Britain is probably not very different from that in their native country (Holmes *et al.*, 1973). Deficiency is most likely to occur in women, rather than in men, since women do not go out so much and eat a less varied diet. Since some food surveys

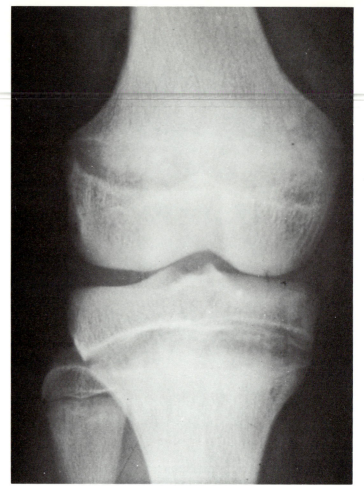

(a)

Figure 4.12. Spontaneous healing of nutritional rickets in an adolescent Asiatic immigrant. The first radiograph (a) was taken on 15.5.1975

have not shown a significantly lower vitamin D intake in infantile and late-onset rickets than in controls, it has been suggested that chapattis counteract the effect of vitamin D, supposedly in the same way as phytate (Ford *et al.*, 1972). However, feeding chapattis does not appear to reduce the increased calcium absorption produced by vitamin D treatment (Dent *et al.*, 1973). In any case phytate is thought to produce insoluble calcium salts within the bowel, reducing the amount available for absorption, which would presumably not cause osteo-malacia. Despite persuasive writing to the contrary (Loomis, 1970), pigmentation of the skin is unlikely to be an important factor in nutritional osteomalacia in the immigrant and, when osteomalacia occurs in the darker West Indian immigrant, it responds well to ultraviolet light (Stamp, 1975).

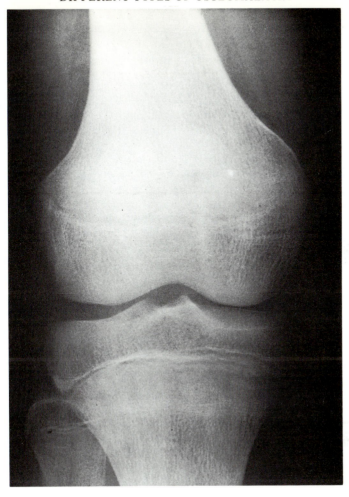

(b)

Figure 4.12. The second radiograph (b) was taken on 8.10.1975

However there is no doubt that in Asian immigrants 25(OH)D levels are low, whatever the cause (Preece *et al.,* 1973) and these appear to increase in the summer, when there is a parallel reversal of hypocalcaemia (Gupta, Round and Stamp, 1974). Rickets may also heal during the summer months (*Figure 4.12*). In immigrants, signs of vitamin D deficiency may appear at any age, with rickets in childhood, with osteomalacia in the adult and a mixture of these two during the rapid growth of adolescence (Cooke *et al.,* 1973). Moncrieff, Lunt and Arthur (1973) also describe nutritional rickets in pubertal Asians. In assessing the incidence of osteomalacia in adolescents it is important to distinguish the biochemical changes from those which normally occur at puberty (Stamp and Round, 1973). Goel *et al.,* (1976) emphasize the continuing evidence of nutritional rickets in Asian children in Glasgow. They found that in such children the

25(OH)D levels were lower than in unaffected children, but that the alkaline phosphatase was of no diagnostic help. Prevention of rickets was suggested by incorporation of a palatable form of vitamin D in the diet, and fortification of chapatti flour with vitamin D has been shown to increase 25(OH)D levels in the serum (Pietrek *et al.*, 1976).

Osteomalacia and malabsorption

Until recently malabsorption due to gluten-sensitive enteropathy (coeliac disease) was probably the commonest cause of osteomalacia in Britain, but this cause has been overtaken in frequency by vitamin D deficiency. Nevertheless, coeliac disease may cause osteomalacia at any age and should always be excluded. Occasionally there are clinical clues such as delayed puberty, or a hypochromic microcytic anaemia unresponsive to iron, but a small intestinal biopsy may be necessary to demonstrate or exclude an atrophic mucosa.

Young people with osteomalacia due to gluten-sensitive enteropathy may show very rapid bone turnover, with very high plasma alkaline phosphatase activity and greatly increased excretion of collagen-derived peptides in the urine. It is difficult to understand the reports of normal true calcium absorption in such patients (Melvin *et al.*, 1970), when the small intestinal mucosa is presumably atrophic. It is not clear why only some patients with coeliac disease develop rickets or osteomalacia, and why it may only become clinically obvious in adult life. The bone disease is presumably the result of poor vitamin D absorption. In personal studies the intravenous administration of $1,25(OH)_2D$ to patients with untreated coeliac disease failed to show any increase in net calcium absorption, an expected result where the small intestinal mucosa was atrophic. This was not the experience of Bordier *et al.*, (1974) using $1\alpha(OH)D$.

Other causes of osteomalacia which come into this group are less common; they include gastric surgery, biliary atresia, biliary cirrhosis and extensive small bowel resection. In the past, partial gastrectomy was considered the commonest cause of osteomalacia in Europe (Thompson, Lewis and Booth, 1966). This is unlikely to be true in Britain today, although the recorded overall incidence of osteomalacia after partial gastrectomy varies widely. Osteomalacia after gastrectomy is probably due to a combination of poor intake and malabsorption of vitamin D (Morgan, Hunt and Paterson, 1970). Reduced exposure to sunlight may be important. Osteomalacia is likely to be less common following vagotomy and pyloroplasty than after the classic partial gastrectomy operations. Nilsson and Westlin (1971) found that the incidence of fractures in patients after gastrectomy was higher than in matched controls. An unexpectedly high incidence of post-gastrectomy osteomalacia of 25 per cent has been recorded by Eddy (1971) from America.

In liver disease the 25(OH)D levels are low, but clinical osteomalacia is not often seen. The cause of these low levels is probably multiple. Skinner *et al.*, (1977) showed that in primary biliary cirrhosis serum 25 (OH)D values become normal if sufficient vitamin D is given parenterally, implying that hepatic 25-hydroxylation was effective; since such patients continue to develop bone disease despite vitamin D, this may be due to factors other than 25-hydroxylation. Krawitt, Grundman and Mawer (1977) demonstrate unimpaired 25-hydroxylation, but poor absorption and excess urinary loss of vitamin D occur.

Osteomalacia and renal disease

Division of the renal causes of osteomalacia into glomerular and tubular is of first importance, and the term 'renal rickets' to cover them should be abolished. The causes of renal-glomerular osteodystrophy are different from those of osteomalacia due to renal-tubular disease, and there are differences in all clinical aspects. With the longer survival of patients with chronic renal failure, the bone disease of renal-glomerular failure is commoner than that due to tubular disease and, at present, is more easily understood.

Renal-glomerular osteodystrophy

Progressive renal failure causes a form of bone disease called renal osteodystrophy (p. 103), which is a mixture of osteomalacia, osteitis fibrosa, osteoporosis, osteosclerosis and subperiosteal new bone formation. Renal failure also causes a number of musculoskeletal problems (Massry *et al.,* 1975).

The dramatic effects of renal failure on the skeleton, particularly in children, remained a curiosity until life could be prolonged by regular haemodialysis (Siddiqui and Kerr, 1971), a procedure which however produced skeletal problems of its own. The advent of haemodialysis and the discoveries about vitamin D have led to considerable advance which it is not possible to give a full account of here; the reader is referred to Stanbury (1972) and Catto (1976) for further details.

Much current work on renal bone disease is contradictory. Symptoms due to it are far less common than histological and biochemical abnormalities, and the geographical incidence and type of disease vary both before dialysis, where the intake of vitamin D is probably important (Lumb, Mawer and Stanbury, 1971), and during dialysis (*Lancet,* 1976). Further, the response to treatment with 1α-hydroxylated vitamin D derivatives is widely variable. However the main points appear clear. First, the work of Stanbury (1957) and Dent, Harper and Philpot (1961) showed that the features of osteomalacia resulting from renal-glomerular failure resembled those of vitamin D deficiency, except that they occurred without such deficiency and required very large doses of vitamin D to alleviate them; second, the discovery of the renal origin of $1,25(OH)_2D$, and its defective formation in renal failure, largely explained the vitamin D resistance in this state, an explanation which was supported by the cure of renal osteodystrophy with microgram quantities of $1,25(OH)_2D$; and third, experimental work showed the importance of small increases in plasma phosphate consequent upon declining renal function, which produced hypocalcaemia and stimulated parathyroid overactivity (Bricker, 1972). Prevention of such increases could reduce parathyroid overactivity.

These points are indicated in *Figure 4.3*. The effect of the metabolic acidosis of renal failure on the skeleton is still disputed and is not included. It has been suggested that in some patients with high bone turnover calcitonin production is relatively decreased (Kanis *et al.,* 1977a); Habener and Schiller, 1977) which could contribute to the excessive bone resorption. This scheme does not explain osteosclerosis and periosteal new bone formation, the causes of which are mysterious.

The proportion of osteomalacia, osteitis fibrosa and osteoporosis before and during haemodialysis differs according to the patients studied (Woods, Bishop and Nicholson, 1972; Ellis and Peart, 1973). Biochemical changes have also been observed in such bones, which include an increase in magnesium content, an increase in the amorphous mineral phase, and a delay in the maturation of collagen.

The combined effects of $1,25(OH)_2D$ deficiency and hyperparathyroidism rapidly destroy the skeleton (*Figure 4.7*). The patient with renal osteodystrophy has all the features of osteomalacia plus those of the underlying disease. There are many causes of renal failure including chronic glomerulonephritis, chronic pyelonephritis and polycystic kidneys. Obstruction by urethral valves (*Figure 4.4*, Smith, 1972) is particularly important in children, because it is potentially reversible. In children severe dwarfism or infantilism may occur and, as in adults, anaemia, pigmentation and general ill-health.

Treatment of renal osteodystrophy includes treatment of the renal failure itself and of the effects such failure produces. Unless the cause of the renal failure can be reversed (for instance by relief of urinary obstruction) it will be variably progressive and in appropriate cases haemodialysis or transplantation will be necessary. The former appears to have little effect on the skeletal changes but may complicate them (p. 123); the latter when successful may reverse the bone disease, except that autonomous hyperparathyroidism may persist.

The effects of renal failure on the skeleton can be ameliorated by prevention or reduction of hyperphosphataemia and by administration of vitamin D or its 1α-hydroxylated derivatives. The plasma phosphate can be reduced by reducing effective phosphate intake with phosphate-binding agents. Although in uraemic dogs control of hyperphosphataemia may delay bone disease (Rutherford *et al.*, 1977) the situation in man is less clear.

Most recent therapeutic interest has been centred around the use of vitamin D metabolites. In patients with renal osteodystrophy 'resistant' to the effect of large doses of vitamin D, it has been demonstrated that $1,25(OH)_2D$ or $1\alpha(OH)D$ given in daily microgram doses increases calcium absorption, corrects hypocalcaemia and improves myopathy (Henderson *et al.*, 1974). When given over a prolonged period such metabolites are capable of healing the osteodystrophy without producing significant side-effects. The changes are more dramatic and predictable in children than in adults. Examples of personally studied patients are cited by Kanis *et al.*, (1977b). The main advantages of the use of $1,25(OH)_2D$ or $1\alpha(OH)D$ compared with the parent vitamin appears to be fewer side effects and shorter half-life (Kanis, Russell and Smith, 1977; Kanis and Russell, 1977) which makes treatment easier to control. Although renal osteodystrophy may respond equally well to large doses of the parent vitamin, intermittent hypercalcaemia, worsening of renal function and ectopic calcification are problems produced by this vitamin which may only be overcome by parathyroidectomy. It is not clear how vitamin D works in renal osteodystrophy since $1,25(OH)_2D$ may be undetectable in such patients. It may be that the block in 1α-hydroxylation is not complete, or that renal osteodystrophy is partly due to a deficiency of $25(OH)D$ (as some have supposed – Eastwood *et al.*, 1976) or that other biologically active metabolites are produced.

The successful treatment of the deformities of renal osteodystrophy requires cooperation between the physician and orthopaedic surgeon. In children progressive knock knee may be halted and osteotomies done if necessary when the bone has healed. In view of the possibility of future renal transplantation it may be desirable to do such operations without blood transfusions.

Bone disease in the dialysed patient.   Although life may be prolonged for many years by haemodialysis, this treatment does not improve the bones and its effects on the skeleton are variable. In Oxford a substantial number of patients beginning dialysis have histological bone disease, but in the next few years only osteoporosis increases significantly in incidence. In some other centres progressive loss of bone (dialysis bone disease) occurs. Its features (Siddiqui and Kerr, 1971) are pain in the feet, groins and shoulders with pathological fractures and progressive periarticular osteoporosis. The alkaline phosphatase is normal, and vitamin D is usually ineffective.

In the dialysed patient treatment with $1\alpha(OH)D$ or $1,25(OH)_2D$ has been less effective than in non-dialysed patients and hypercalcaemia has been a problem. In part this ineffectiveness has been attributed to phosphate deficiency produced by dialysis. The severity of renal osteodystrophy may be reduced by dialysis against a high calcium fluid and by oral calcium supplementation.

Other skeletal features in chronic renal failure.   Apart from renal osteodystrophy, patients with renal failure may develop aseptic necrosis of bone particularly in the femoral head. This tends to occur after transplantation, and is likely to be due to the added effects of corticosteroid therapy.

Non-skeletal features in chronic renal failure.   One of the most important non-skeletal effects of renal failure is metastatic calcification, which may occur in the periarticular soft tissues, the blood vessels, the eyes, the skin and elsewhere. Whilst the effects on the blood vessels are clinically the most significant, periarticular calcification is the form of metastatic calcification most easily corrected either by reduction in phosphate, by parathyroidectomy or by reduction of vitamin D dosage (*see also* Chapter 10). Patients on haemodialysis may also develop symptoms due to crystal-induced arthritis (Massry *et al.*, 1975).

Renal-tubular osteomalacia

There are very many renal-tubular causes of osteomalacia, all of them uncommon, and several very rare. However, it is important to be aware of them and to emphasize again how they differ from renal-glomerular osteodystrophy. The development of knowledge about them is well described by Dent (1970) and Fanconi and Prader (1972).

Hypophosphataemic rickets.   This is the commonest form of renal-tubular rickets and also known as vitamin-D-resistant rickets (VDRR) or Type I rickets (Dent, 1952). The basic defect in this condition appears to be an inherited abnormality of phosphate transport by the renal tubule and by other transporting tissues, notably the gut. The 'clearance' of phosphate is increased, the maximum tubular reabsorptive capacity for phosphate is reduced and the TmP/GFR (Chapter 2) is lower than normal. This defect is inherited as an X-linked dominant. Only some subjects with this trait develop rickets, and it tends to be more severe in males, presumably due to the unopposed action of the affected X chromosome.

Since hypophosphataemia is a constant feature, it has been suggested that this is due to secondary hyperparathyroidism, but there is very little evidence of this and no convincing reason to suggest it. Thus PTH levels are often (Arnaud, Glorieux and Scriver, 1971), but not always (Reitz and Weinstein, 1973), normal; neither histology nor x-rays show excessive bone resorption; and there is no cause for the supposed parathyroid overactivity, the plasma calcium being normal.

The term vitamin-D-resistant rickets also implies that the condition has something to do with vitamin D, but again there is no direct evidence for this (Russell *et al.*, 1975). The fact that vitamin D in large doses may partly cure the rickets may be due to its pharmacological effects in increasing calcium and phosphate absorption.

Importantly, this condition is clinically unlike either nutritional osteomalacia or that occasionally induced by severe phosphate deficiency due to excessive administration of aluminium hydroxide.

The features of Type 1 rickets may begin in infancy although hypophosphataemia may be difficult to confirm at this age unless the control values are adequate. Rickets presents as deformity, and the child is otherwise well, without muscle weakness. Later such children may walk back and forth to school without difficulty despite obvious deformity of the limbs (knock knee or bow legs). The child (and later the adult) is dwarfed, often with disproportionately short limbs but a near normal sitting height. Growth is usually below the third percentile and it is unusual to reach 5 ft (147 cm) in height. Menarche may be early. The shape of the face and skull is said to be characteristic, often with a prominent forehead and long anteroposterior measurement.

The family history is often helpful; the only biochemical abnormality is hypophosphataemia; plasma and urine calcium are normal and there is no aminoaciduria (Table 4.II). The plasma alkaline phosphatase may be normal for age. X-rays may show severe rickets. Later the bones are often dense, and the long bones buttressed. Exostoses occur, and Looser's zones may be seen.

The rickets appear to improve radiologically with high doses of vitamin D (up to 2 mg daily) and the plasma alkaline phosphatase may fall. There is however a risk of vitamin D intoxication and under-treatment is preferable. Vitamin D may produce no histological change in the bones but data on this are scanty. However McNair and Stickler (1969) in a review of 36 patients could find no evidence that vitamin D alone altered eventual height, or that deformity or plasma phosphate level was altered by this treatment. In comparison to the adult-onset non-familial disease (p. 125) the use of phosphate alone in this condition is not yet established, and there is a practical difficulty in maintaining a normal level of phosphate in the serum. However, since the use of phosphate may reduce the requirement of vitamin D, appears to increase linear growth (Glorieux *et al.*, 1972), and also appears to be harmless, there is much to be said for giving it with vitamin D. To be effective phosphate has to be given every four hours five times daily, in a total dose of 1 to 4 g of inorganic phosphate per day. There seems to be no advantage in using vitamin D metabolites in this disorder.

Apart from the complications of treatment, it is known that patients with vitamin-D-resistant rickets may develop spinal cord compression due to ossification of the ligamenta flava (Highman, Sanderson and Sutcliffe, 1970). This is shown in *Figure 4.10 (b)*.

It is of interest to compare the clinical features of VDRR with those of pure phosphate depletion induced by excessive aluminium hydroxide ingestion (Lotz, Zisman and Bartter, 1968). This may produce a form of hypophosphataemic osteomalacia with malaise, anorexia and muscle weakness (Lotz, Ney and Bartter, 1964), together with hypercalcuria and increased intestinal absorption of calcium. It is possible that the low phosphate stimulates $1,25(OH)_2D$ production.

In contrast the normal recorded levels of $1,25(OH)_2D$ in VDRR suggest that in this disorder the intracellular concentration of phosphate may be normal.

Adult-onset hypophosphataemic osteomalacia. This is rare, but of considerable interest. In this condition, which is not inherited, the previously well adolescent (Fanconi, 1971) or young adult rapidly develops the bone pain and proximal weakness of severe osteomalacia and is found to have hypophosphataemia. There may be an excess of glycine in the urine. Radiology and histology confirm the diagnosis and the patient responds well to phosphate alone (Nagant de Deuxchaisnes and Krane, 1967) or vitamin D and phosphate (Smith and Dick, 1968; Dent and Stamp, 1971). The cause of this condition is unknown, but amongst this group are the patients with non-endocrine tumours of the soft tissues (p. 126). It is important that in any adult developing hypophosphataemic osteomalacia, a clinical and radiological search should be made for such a tumour. Hypophosphataemia has also been described in neurofibromatosis (Dent, 1976) and in fibrous dysplasia, where it can be reversed by removal of the abnormal bony tissue (Dent and Gertner, 1976).

Other renal-tubular syndromes. All other tubular osteomalacic syndromes are rare (Dent, 1970; Fanconi and Prader, 1972). The early classifications of Dent (1952) and Dent and Harris (1956) are still very useful, and deal with the disorders in increasing complexity of the renal-tubular lesion. Thus osteomalacia may be associated with an inability of the renal tubule to acidify the urine or to reabsorb bicarbonate. This causes a hyperchloraemic acidosis and the osteomalacia responds to vitamin D and alkali or alkali alone (Richards, Chamberlain and Wrong, 1972). The acidosis may be acquired, as after ureterosigmoid anastomosis.

Osteomalacia (and rickets) may result from multiple renal-tubule defects which are included under the general heading of the 'Fanconi syndrome'. The common feature of this syndrome is generalized amino-aciduria. In children the commonest cause of the Fanconi syndrome is nephropathic cystinosis, in which massive deposits of cystine crystals accumulate in the tissues and can be seen in the eyes (Chapter 2). The condition is presumed to be a recessively-inherited inborn error of metabolism but the exact defect has not been identified. Characteristically the affected children fail to thrive towards the end of the first year of life, with rapid weight loss and dehydration due to failure of renal reabsorption of water. Rickets develops early (*Figure 4.6*). Its causes may be multiple but it responds to phosphate together with vitamin D or a 1α-hydroxylated metabolite (Etches, Pickering and Smith, 1977). Death from progressive renal failure usually occurs in late childhood.

Rarer causes of childhood Fanconi syndrome include the X-linked oculocerebrorenal or Lowe's syndrome; and familial occurrence in children who do not have cystinosis is described (Sheldon, Luder and Webb, 1961). Dominantly-inherited adult-onset Fanconi syndrome has also been described (Smith, Lindenbaum and Walton (1976). Generalized amino-aciduria is also a feature of Wilson's disease, cadmium poisoning, poisoning with other heavy metals, and myeloma (where the cause is not clear). The osteomalacia (or rickets) in most patients with the Fanconi syndrome responds to vitamin D, but for details of treatment individual papers should be consulted.

There are three causes of rickets which do not fit into the main categories we have described. These are the rickets associated with anticonvulsant therapy, with non-endocrine tumours, and that presumed to be due to a deficiency of the renal l-hydroxylase, vitamin-D-dependent rickets.

Anticonvulsant osteomalacia

There is a higher than normal incidence of rickets and osteomalacia in patients on anticonvulsants (Anast, 1975). Initial studies were concerned with residents of epileptic colonies but later out-patients were examined (Hahn et al., 1975). Epileptics in institutions might be expected to be deficient in vitamin D because of a relatively poor diet and poor exposure to sunlight. However it does seem that other factors are important. Phenobarbitone and phenytoin reduce the hormonal activity of a wide range of administered steroids, probably by hepatic enzyme induction, and it is thought that vitamin D is converted to hydroxylated derivatives which are not biologically active (Dent et al., 1970). If this is so the requirement for vitamin D would be increased. Osteomalacia in patients on anticonvulsants has been shown to have a moderate resistance to vitamin D and a rapid response to microgram quantities of 25(OH)D (Stamp et al., 1972). The effects of more than one anticonvulsant appear to be cumulative. Thus Hahn et al., (1975) found a greater decrease in serum calcium, serum 25(OH)D and bone mass in his rachitic children on phenobarbitone plus phenytoin, than on either anticonvulsant alone. Osteomalacia has also been reported after prolonged glutethimide administration (Greenwood, Prunty and Silver, 1973).

Tumour rickets

Rickets or osteomalacia may occur in association with a variety of tumours, particularly sclerosing haemangiomata (Salassa, Jowsey and Arnaud, 1970) and non-ossifying fibroma (Pollack, Schiller and Crawford, 1973), but also giant cell tumours. Removal of the tumours has resulted in recovery from osteomalacia. The osteomalacia is of the hypophosphataemic variety, with considerable bone disease and myopathy, and is a significant cause of acquired hypophosphataemic osteomalacia in adults (p. 125). Since removal of the tumour cures the bone disease, it is reasonable to postulate that the tumour produces some hormonal substance, but there are no clues as to what this might be. Suggestions have been made that $1,25(OH)_2D$ production is defective but the clinical picture does not fit with this. Where the tumours cannot be completely removed treatment of the hypophosphataemia with phosphate alone or with phosphate plus vitamin D would seem to be logical.

Vitamin-D-dependent rickets

In this rare form of rickets (Dent, Friedman and Watson, 1968; Scriver, 1970) the features of severe nutritional rickets are associated with an increased requirement for vitamin D. It does not resemble Type 1 vitamin-D-resistant rickets, since severe bone pain, tenderness, and marked proximal muscle weakness are features, and since there is hypocalcaemia in addition to hypophosphataemia. The x-rays (*Figure 4.5*) are similar to those of severe nutritional rickets and deformity and dwarfism may be severe (Smith, 1972). Inheritance is recessive. Fraser *et al.,* (1973) demonstrated that the rickets in patients with this disorder responded only to large doses of vitamin D or 25(OH)D but that daily microgram doses of 1,25(OH)$_2$D given to one child produced rapid biochemical and radiological improvement. These results suggest that vitamin-D-dependent rickets is

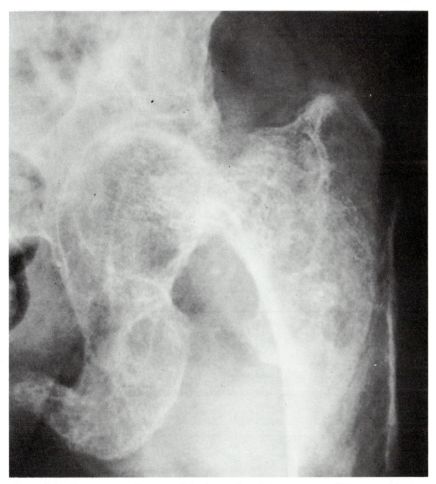

Figure 4.13. Spontaneous fracture of the femoral neck in a woman of 70 years with osteomalacia. Both femoral necks were fractured and she had been unable to walk for some time. A diagnosis of Paget's disease had been made because of the raised alkaline phosphatase. She was on anticonvulsants. On vitamin D she rapidly improved and the fractures united without surgery

due to a block in the formation of 1,25 $(OH)_2D$ from 25 $(OH)D$, presumably due to defective action of the renal-l-hydroxylase. Since the condition can be cured by native vitamin D in adequate dose, this is probably still the appropriate treatment. Such high doses presumably act by producing sufficient 1,25 $(OH)_2D$ despite the defective enzyme. A consideration of the possible biochemical causes of vitamin-responsive inborn errors of metabolism is provided by Scriver (1973).

DISCUSSION

The rapid biochemical advance in knowledge of vitamin D metabolism has led to the development of useful assays for its circulating metabolites, and has certainly helped us to understand the various causes of osteomalacia. There are still mysteries left, particularly in the field of renal bone disease. The causes of tumour rickets and Type 1 vitamin-D-resistant rickets also remain obscure, and more investigation is necessary on the effect of vitamin D on its known (and unknown) target organs. The concentration of research on the undoubtedly important renal hormone $1,25(OH)_2D$ should not exclude the possibility of other biologically-active metabolites. There are still some surprises left in the vitamin D field. One example of this is the demonstration that most vitamin D in human milk is in the aqueous, rather than the lipid, phase as a water-soluble conjugate with sulphate (Lakdawala and Widdowson, 1977).

Clinically it remains important to diagnose osteomalacia as soon as possible, since it is a readily curable cause of disability (*Figure 4.13*). There is a widely held belief that the diagnosis of the different types of osteomalacia is too complex to attempt; but this is not so, and such distinctions are therapeutically important.

SUMMARY

In rickets and osteomalacia the cardinal abnormality is defective mineralization of bone and the main cause a lack of vitamin D or a disturbance of its metabolism. Thus the recent discoveries on this vitamin, now regarded as a hormone, have been of considerable clinical help. Although osteomalacia is not common, it is disabling and responds well to treatment. For these reasons the diagnosis should not be missed.

REFERENCES

Aaron, J. E., Gallagher, J. C., Anderson, J., Stasiak, L., Longton, E. B., Nordin, B. E. C. and Nicholson, M. (1974a). Frequency of osteomalacia and osteoporosis in fractures of the proximal femur. *Lancet* 1, 229–233

Aaron, J. E., Gallagher, J. C. and Nordin, B. E. C. (1974b). Seasonal variation of histological osteomalacia in femoral-neck fractures. *Lancet* 2, 84–85

Albright, F. and Reifenstein, E. C. (1948). *The Parathyroid Glands and Metabolic Bone Disease*. Baltimore: Williams and Wilkins

Anast. C. S. (1975). Anticonvulsant drugs and calcium metabolism. *New Engl. J. Med.* 292, 587–588

Arnaud, C., Glorieux, F. and Scriver, C. (1971). Serum parathyroid hormone in X-linked hypophosphataemia. *Science, N. Y.* 173, 845–847

REFERENCES

Birge, S. J., and Haddad, J. G. (1975). 25-hydroxycholecalciferol stimulation of muscle metabolism. *J. clin. Invest.* **56**, 1100–1107

Bordier, P., Pechet, M., Hesse, R., Marie, P. and Rasmussen, H. (1974). Response of adult patients with osteomalacia to treatment with crystalline 1 α-hydroxy vitamin D_3. *New Engl. J. Med.* **291**, 866–871

Bordier, Ph. J. and Tun Chot, S. (1972). Quantitative histology of metabolic bone disease. *Clinics Endocr. Metabolism* **1**, 197–215

Bricker, N. S. (1972). On the pathogenesis of the uraemic state. An exposition of the 'Trade-off hypothesis'. *New Engl. J. Med.* **286**, 1093–1099

Catto, G. R. D. (1976). Renal bone disease. *Jl. R. Coll. Physns.* **11**, 75–85

Chalmers, J. (1970). Subtrochanteric fractures in osteomalacia. *J. Bone Jt Surg.* **52B**, 509–513

Cooke, W. T., Swan, C. H. J., Asquith, P., Melikian, V. and McFeely, W. E. (1973). Serum alkaline phosphatase and rickets in urban schoolchildren. *Br. med. J.* **1**, 324–327

Curry, O. B., Basten, J. F., Francis, M. J. O. and Smith, R. (1974). Calcium uptake by sarcoplasmic reticulum of muscle from vitamin D deficient rabbits. *Nature* **249**, 83–84

Dent, C. E. (1952). Rickets and osteomalacia from renal tubule defects. *J. Bone Jt Surg.* **34B**, 266–274

Dent, C. E. (1970). Rickets (and osteomalacia): nutritional and metabolic (1919–1969). *Proc. R. Soc. Med.* **63**, 401–408

Dent, C. E. (1976). Metabolic forms of rickets (and osteomalacia). In *Inborn Errors of Calcium and Bone Metabolism*. Eds. H. Bickel and J. Stern. Lancaster: MTP Press

Dent, C. E. and Gertner, J. M. (1976). Hypophosphataemic osteomalacia in fibrous dysplasia. *Q. Jl Med.* **45**, 411–420

Dent, C. E. and Harris, H. (1956). Hereditary forms of rickets and osteomalacia. *J. Bone Jt Surg.* **38B**, 204–226

Dent, C. E. and Stamp, T. C. B. (1971). Hypophosphataemic osteomalacia presenting in adults. *Q. Jl Med.* **40**, 303–329

Dent, C. E. and Stamp, T. C. B. (1977). Vitamin D, rickets and osteomalacia. In *Metabolic Bone Disease*. Eds L. V. Avioli and S. M. Krane. Vol. 1. pp. 237–305. New York and London: Academic Press

Dent, C. E., Friedman, M. and Watson, L. (1968). Hereditary pseudo-vitamin D deficiency rickets *J. Bone Jt Surg.* **50B**, 708–719

Dent, C. E., Harper, C. M. and Philpot, G. K. (1961). The treatment of renal-glomerular osteodystrophy. *Q. Jl Med.* **30**, 1–32

Dent, C. E., Richens, A., Rowe, D. J. F. and Stamp, T. C. B. (1970). Osteomalacia with long-term anticonvulsant therapy in epilepsy. *Br. med. J.* **4**, 69–72

Dent, C. E., Round, J. M., Rowe, D. J. F. and Stamp, T. C. B. (1973). Effect of chapattis and ultraviolet irradiation on nutritional rickets in an Indian immigrant. *Lancet* **1**, 1282–1284

Eastwood, J. B., Harris, E, Stamp, T. C. B. and de Wardener, H. E. (1976). Vitamin D deficiency in the osteomalacia of renal failure. *Lancet* **2**, 1322–1326

Eddy, R. L. (1971). Metabolic bone disease after gastrectomy. *Am. J. Med.* **50**, 442–448

Ellis, H. A. and Peart, K. M. (1973). Azotaemic renal osteodystrophy: a quantitative study on iliac bone. *J. clin. Path.* **26**, 83–101

Emmerson, B. T. (1970). 'Ouch-Ouch' disease: the osteomalacia of cadmium nephropathy. *Ann. intern. Med.* **73**, 854–855

Etches, P., Pickering, D. and Smith, R. (1977). Cystinotic rickets treated with vitamin D metabolites. *Archs. Dis. Childh.* **52**, 661–664

Fanconi, A. (1971). Idiopathische hypophosphatamische Osteomalazie mit beginn in der Adoleszenz. *Helv. paediat. Acta* **26**, 535–549

Fanconi, A, and Prader, A. (1972). Hereditare rachitis formen. *Schweiz. med. Wschr.* **102**, 1073–1078

Ford, J. A., Colhoun, E. M., McIntosh, W. B. and Dunnigan, M. G. (1972). Rickets and osteomalacia in the Glasgow Pakistan community, 1961–1971. *Br. med. J.* **2**, 677–680

Fraser, D., Kooh, S. W., Kind, H. P., Holick, M. F., Tanaka, Y. and DeLuca, H. F. (1973). Pathogenesis of vitamin-D-dependent rickets. *New Engl. J. Med.* **289**, 817–822

Glorieux, F. H., Scriver, C. R., Reade, T. M., Goldman H. and Roseborough A. (1972). Use of phosphate and vitamin D to prevent dwarfism and rickets in X-linked hypophosphataemia. *New Engl. J. Med.* **287**, 481–487

REFERENCES

Goel, K. M., Sweet, E. M., Logan, R. W., Warren, J. M., Arneil, G. C. and Shanks, R. A. (1976). Florid and subclinical rickets among immigrant children in Glasgow. *Lancet* **1**, 1141–1148

Greenwood, R. H., Prunty, F. T. G. and Silver, J. (1973). Osteomalacia after prolonged glutethimide administration. *Br. med. J.* **1**, 643–645

Gupta, M. M., Round, J. M. and Stamp, T. C. B. (1974). Spontaneous cure of vitamin-D deficiency in Asians during summer in Britain. *Lancet* **1**, 586–588

Habener, J. F. and Schiller, A. L. (1977). Pathogenesis of renal osteodystrophy–a role for calcitonin? *New Engl. J. Med.* **296**, 112–113

Hahn, T. J., Hendin, B. A., Scharp, C. R., Boisseau, V. C. and Haddad, J. G. (1975). Serum 25-hydroxycholecalciferol levels and bone mass in children on chronic anticonvulsant therapy. *New Engl. J. Med.* **292**, 550–554

Haussler, M. R. and McCain, T. A. (1977). Basic and clinical concepts related to vitamin D metabolism and action. *New Engl. J. Med.* **297**, 974–983, 1041–1050

Henderson, R. G., Russell, R. G. G., Ledingham, J. G. G., Smith, R., Oliver, D. O., Walton, R. J., Small, D. G., Preston, C., Warner, C. T. and Norman, A. W. (1974). Effects of 1,25-dihydroxycholecalciferol on calcium absorption, muscle weakness and bone disease in chronic renal failure. *Lancet* **1**, 379–384

Highman, J. H., Sanderson, P. H. and Sutcliffe, M. M. L. (1970). Vitamin-D-resistant osteomalacia as a cause of cord compression. *Q. Jl Med.* **39**, 529–537

Hodkinson, H. M., Round, P, Stanton, B. R. and Morgan, C. (1973). Sunlight, vitamin D and osteomalacia in the elderly. *Lancet* **1**, 910–912

Holmes, A. M., Enoch, B. A., Taylor, J. L. and Jones, M. E. (1973). Occult rickets and osteomalacia amongst the Asian immigrant population. *Q. Jl Med.* **42**, 125–149

Jenkins, D. H. R., Roberts, J. G., Webster, D. and Williams, E. O. (1973). Osteomalacia in elderly patients with fracture of the femoral neck. *J. Bone Jt Surg.* **55B**, 575–580

Kanis, J. A. and Russell, R. G. G. (1977). Rate of reversal of hypercalcaemia and hypercalcuria induced by vitamin D and its 1 α-hydroxylated derivatives. *Br. med. J.* **1**, 78–81

Kanis, J. A., Russell, R. G. G. and Smith, R. (1977). Physiological and therapeutic differences between vitamin D, its metabolites and analogues. *Clin. Endocr.* 7, Suppl. 191s–201s

Kanis, J. A., Earnshaw, M., Heynen, G., Ledingham, J. G. G., Oliver, D. O., Russell, R. G. G., Woods, C. G. Franchimont, P. and Gaspar, S. (1977a). Bone turnover rates after bilateral nephrectomy: role of calcitonin. *New Engl. J. Med.* **296**, 1073

Kanis, J. A., Henderson, R. G., Heynen, G., Ledingham, J. G. G., Russell, R. G. G., Smith, R and Walton, R. J. (1977b). Renal osteodystrophy in non-dialysed adolescents. Long-term treatment with 1 α-hydroxycholecalciferol. *Archs Dis. Childh.* **52**, 473–481

Kodicek, E. (1974). The story of vitamin D. From vitamin to hormone. *Lancet* **1**, 325–329

Krawitt, E. L., Grundman, M. J. and Mawer, E. B. (1977). Absorption, hydroxylation and excretion of vitamin D_3 in primary biliary cirrhosis. *Lancet* **2**, 1246–1249

Lakdawala, D. R. and Widdowson, E. (1977). Vitamin D in human milk. *Lancet* **1**, 167–168

Lancet (1976). Dialysis osteodystrophy. (Editorial). **2**, 451–452

Loomis, W. F. (1970). Rickets. *Scient. Am.* **223**, 77–91

Lotz, M., Ney, R. and Bartter, F. C. (1964). Osteomalacia and debility resulting from phosphorus depletion. *Trans. Ass. Am. Physns.* 77, 281–295

Lotz, M., Zisman, E. and Bartter, F. C. (1968). Evidence for a phosphorus-depletion syndrome in man. *New Engl. J. Med.* **278**, 409–415

Lumb, G. A., Mawer, E. B. and Stanbury, S. W. (1971). The apparent vitamin D resistance of chronic renal failure. *Am. J. Med.* **50**, 421–441

Lund, B., Sorensen, O. H. and Christensen, A. B. (1975). 25-hydroxycholecalciferol and fractures of the proximal femur. *Lancet* **2**, 300–302

Maddison, P. J. and Bacon, P. A. (1974). Vitamin-D-deficiency, spontaneous fractures and osteopenia in rheumatoid arthritis. *Br. med. J.* **4**, 433–435

McLaughlin, M., Raggatt, P. R., Fairney, A., Brown, D. J., Lester, E. V. and Wills, M. R. (1974). Seasonal variation in serum 25-hydroxycholecalciferol in healthy people. *Lancet* **1**, 536–538

McNair, S. L. and Stickler, G. B. (1969). Growth in familial hypophosphataemic vitamin-D-resistant rickets. *New Engl. J. Med.* **281**, 511–516

Massry, S. G., Bluestone, R., Klinenberg, J. R. and Coburn, J. W. (1975). Abnormalities of the musculo-skeletal system in haemodialysis patients. *Semin. Arthritis Rheumatism* **4**, 321–349

REFERENCES

Melvin, K. E. W., Hepner, G. W., Bordier, P., Neale, G. and Joplin, G. F. (1970). Calcium metabolism and bone pathology in adult coeliac disease. *Q. Jl Med.* **39**, 83–113

Moncrieff, M. W., Lunt, H. R. W. and Arthur, L. J. H. (1973). Nutritional rickets at puberty. *Archs. Dis. Childh.* **48**, 221–224

Morgan, B. and Fourman, P. (1969). The diagnosis of osteomalacia and osteoporosis. *Br. J. hosp. Med.* **2**, 901–908

Morgan, D. B., Hunt, G. and Paterson, C. R. (1970). The osteomalacia syndrome after stomach operations. *Q. Jl Med.* **39**, 395–410

Nagant de Deuxchaisnes, C. and Krane, S. M. (1967). The treatment of adult phosphate diabetes and Fanconi syndrome with neutral sodium phosphate. *Am. J. Med.* **43**, 508–543

Nelson, W. E., Vaughan, V. C. and McKay, R. J. (Eds) (1969). *Textbook of Paediatrics* 9th edition. Philadelphia: W. B. Saunders Co

Nilsson, B. E. and Westlin, N. E. (1971). The fracture incidence after gastrectomy. *Acta chir. scand.* **137**, 533–534

O'Driscoll, M. (1973). Subcapital fracture types and osteomalacia and vitamin D deficiency. *J. Bone Jt Surg.* **55B**, 882

Parfitt, A. M. and Frame, B. (1972). Treatment of rickets and osteomalacia. *Semins Drug Treatment.* **2**, 83–115

Pietrek, J., Preece, M. A., Winds, J., O'Riordan, J. L. H., Dunnigan, M. G., McIntosh, W. S. O. and Ford, J. A. (1976). Prevention of Vitamin D deficiency in Asians. *Lancet* **1**, 1145–1148

Pollack, J. A., Schiller, A. L. and Crawford, J. D. (1973). Rickets caused by non-ossifying fibroma. *Pediatrics,* **52**, 364–371

Preece, M. A., McIntosh, W. B., Tomlinson, S., Ford, J. A., Dunnigan, M. G. and O'Riordan, J. L. H. (1973). Vitamin D deficiency among Asian immigrants to Britain. *Lancet* **1**, 907–910

Preece, M. A., Tomlinson, S., Ribot, C. A., Pietrek, J., Korn, H. T., Davies, D. M., Ford, J. A., Dunnigan, M. G. and O'Riordan, J. L. H. (1975). Studies of vitamin D deficiency in man. *Q. Jl Med.* **XLIV**, 575–589

Reitz, R. E. and Weinstein, R. L. (1973). Parathyroid hormone secretion in familial vitamin-D-resistant rickets. *New Engl. J. Med.* **289**, 941–945

Richards, P., Chamberlain, M. J. and Wrong, O. M. (1972). Treatment of osteomalacia of renal-tubular acidosis by sodium bicarbonate alone. *Lancet* **2**, 994–999

Round, J. M. (1973). Plasma calcium, magnesium, phosphorus and alkaline phosphatase levels in normal British schoolchildren. *Br. med. J.* **3**, 137–140

Russell, R. G. G., Smith, R., Preston, C., Walton, R. J., Woods, C. G. Henderson, R. G. and Norman, A. W. (1975). The effect of 1,25-dihydroxycholecalciferol on renal-tubular reabsorption of phosphate, intestinal absorption of calcium and bone histology in hypo-phosphataemic renal tubular rickets. *Clin. Sci. molec. Med.* **48**, 177–186

Rutherford, W. E., Bordier, P., Marie, P., Hruska, K., Harter, H., Greenwalt, A., Blondin, J., Haddad, J., Bricket, N. and Slatopolsky, E. (1977). Phosphate control and 25-hydroxy-cholecalciferol administration in preventing experimental renal osteodystrophy in the dog. *J. Clin. Invest.* **60**, 332–341

Salassa, R. M., Jowsey, J. and Arnaud, C. E. (1970). Hypophosphataemic osteomalacia associated with 'non-endocrine' tumours. *New Engl. J. Med.* **283**, 65–70

Schott, G. D. and Willis, M. R. (1976). Muscle weakness in osteomalacia. *Lancet* **1**, 626–628

Scriver, C. R. (1970). Vitamin D dependency. *Pediatrics* **45**, 361–363

Scriver, C. R. (1973). Vitamin-responsive inborn errors of metabolism. *Metabolism,* **22**, 1319–1344

Sheldon, W., Luder, J. and Webb, B. (1961). A familial tubular absorption defect of glucose and amino-acids. *Archs Dis. Childh.* **36**, 90–95

Siddiqui, J. and Kerr, D. N. S. (1971). Complications of renal failure and their response to dialysis. *Br. med. Bull.* **27**, 153–159

Skinner, R. K., Long, R. G., Sherlock, S. and Wills, M. R. (1977). 25-hydroxylation of vitamin D in primary biliary cirrhosis, *Lancet* **1**, 720–721

Smith, R. (1972). The pathophysiology and management of rickets. *Orthop. Clinics N. Am.* **3**, 601–621

Smith, R. (1976). Bone disease in the elderly. *Proc. R. Soc. Med.* **69**, 925–926

131

REFERENCES

Smith, R. and Dick, M. (1968). The effect of vitamin D and phosphate on urinary total hydroxyproline excretion in adult presenting vitamin D type I renal-tubular osteomalacia. *Clin. Sci.* **35**, 575–587

Smith, R. and Stern, G. (1967). Myopathy, osteomalacia and hyperparathyroidism. *Brain.* **90**, 593–602

Smith, R., Lindenbaum, R. H. and Walton, R. J. (1976). Hypophosphataemic osteomalacia and Fanconi syndrome of adult-onset with dominant inheritance. *Q. Jl Med.* **45**, 387–400

Stamp, T. C. B. (1973). Vitamin D metabolism. Recent advances. *Archs. Dis. Childh.* **48**, 2–7

Stamp, T. C. B. (1975). Factors in human vitamin D nutrition and in the production and cure of classical rickets. *Proc. Nutr. Soc,* **34**, 119–130

Stamp, T. C. B. and Round, J. M. (1973). Serum alkaline phosphatase and rickets. *Br. med. J.* **2**, 113

Stamp, T. C. B., Haddad, J. G. and Twigg, C. A. (1977). Comparison of oral 25-hydroxy-cholecalciferol, vitamin D and ultraviolet lights as determinants of circulating 25-hydroxy vitamin D. *Lancet* **1**, 1341–1343

Stamp, T. C. B., Round, J. M., Rowe, D. J. F. and Haddad, J. G. (1972). Plasma levels and therapeutic effect of 25-hydroxycholecalciferol in epileptic patients taking anticonvulsant drugs. *Br. med. J.* **4**, 9–12

Stanbury, S. W. (1957). Azotaemic renal osteodystrophy. *Br. med. Bull.* **13**, 57–60

Stanbury, S. W. (1972). Azotaemic renal osteodystrophy. *Clinics Endocr. Metabolism* **1**, 267–304

Thompson, G. R., Lewis, B. and Booth, C. C. (1966). Vitamin-D absorption after partial gastrectomy. *Lancet* **1**, 457–485

Woods, C. G., Bishop, M. C. and Nicholson, G. D. (1972). Bone histological changes occurring after haemodialysis treatment for chronic renal failure. *J. Path.* **107**, 137–143

5

Paget's Disease of Bone

INTRODUCTION

The pathological hallmark of this common disorder is excessive and disorganized activity of bone which involves both matrix and mineral. We have no idea of its cause and little knowledge of its natural history, but the biochemical changes produced by the overactive bone justify its inclusion as a metabolic bone disease. It is the control of these abnormalities by agents which appear to act directly on bone that accounts for much of the recent interest in Paget's disease. Since medical treatment can now suppress the overactivity of the bone we can no longer so easily ignore Paget's disease, but it will take many years to decide whether such treatment can prevent its complications. The reader should remember that the results of many therapeutic studies are influenced by enthusiasm as well as by science.

It is now more than a century since Paget's description of this disease, and it undoubtedly existed for many years before. Comprehensive descriptions of this disease are given by Nagant de Deuxchaisnes and Krane (1964), Barry (1969), and Singer (1977).

PATHOPHYSIOLOGY

Since the cause of this disorder is obscure, much of what we see in a Pagetic bone may be wrongly interpreted. With that proviso the main abnormality appears to be one of disorganized overactivity. Bone is being rapidly resorbed and new bone equally rapidly laid down. These activities are closely linked. Histological sections show increased numbers of osteoblasts and osteoclasts and fibrosis within the bone marrow.

The collagen of the bone matrix is not laid down in its normal regular fashion (possibly because of the speed of its deposition) and birefringence is therefore reduced. Because new bone formation is increased, the percentage of bone surface covered with osteoid is increased, and mineralization may not keep up with matrix formation, increasing osteoid thickness. The tidemarks of this disorderly activity and of cycles of resorption and formation produce the wavy cement lines (*Figure 5.1*). These activities may also be expressed in quantitative terms (Table 5.I), which also demonstrate a marked decrease in the number of inert trabecular surfaces, and an increase in the ratio of total resorption to total deposition surfaces. In this group of patients from Oxford (Table 5.I) marrow fibrosis was seen only in Pagetic bone.

TABLE 5.I

Quantitative bone histology on trabecular bone from transiliac bone biopsies.
All patients had radiological evidence of Paget's disease, but in some of these
the biopsies did not include Pagetic bone (Walton *et al.*, 1977b).

Measurement	Pagetic	Non-Pagetic	t	P
Bone area (% total section)	39.7 ± 3.4 (26)	19.4 ± 2.5 (15)	4.27	<0.001
Total deposition surfaces (% total trabecular surfaces)	71.7 ± 2.5 (27)	60.0 ± 3.5 (16)	2.86	<0.01
Total resorption surfaces (% total trabecular surfaces)	26.5 ± 2.1 (27)	6.9 ± 2.4 (17)	6.08	<0.001
Total resorption surfaces + total deposition surfaces	0.41 ± 0.05 (27)	0.13 ± 0.05 (27)	3.94	<0.001
Osteoid area (% total bone matrix)	8.2 ± 1.0 (26)	3.5 ± 0.5 (15)	3.60	<0.001
Osteoid width (area/total deposition surfaces)	0.12 ± 0.01 (26)	0.06 ± 0.01 (15)	3.01	<0.01
Active deposition surfaces (% total deposition surfaces)	34.3 ± 4.0 (27)	2.2 ± 0.8 (17)	6.43	<0.001
Active resorption surfaces (% total resorption surfaces)	20.0 ± 2.0 (27)	2.9 ± 1.9 (17)	6.01	<0.001

The numbers in brackets refer to the number of biopsies. Results are given as mean ± SEM.
All the measurements of bone activity are considerably and significantly increased in Pagetic
bone compared with non-Pagetic bone.

The busy bone of Paget's disease is very vascular; this excessive vascularity is generally regarded as result, rather than as a cause, of the bone disorder. Macroscopically the Pagetic bone is deformed and characteristically becomes bigger than normal. However, the enlarged bone is structurally weak and the long bones tend to crack, often in many places, along their convexity. The bony enlargement may compress nerves, and the prolonged cellular over-activity may predispose to sarcomata.

The overactivity of the bone causes considerable biochemical abnormality. The well-known increase in alkaline phosphatase, considered a measure of osteoblastic activity, appears to be proportional to the radiological extent of the disease (Khairi *et al.*, 1973; Franck *et al.*, 1974). Its concentration varies slowly and unpredictably from year to year with a tendency to increase with

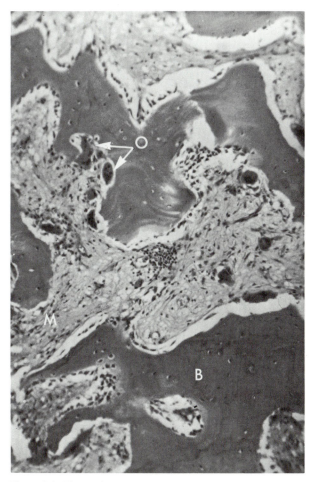

Figure 5.1. The qualitative appearance of Pagetic bone. There is considerable cellular activity, numerous multinucleated osteoclasts, fibrosis of the bone marrow and a mosaic pattern in the mineralized bone. (Decalcified paraffin section. H & E stain × 250. O = osteoclast, M = marrow, B = bone)

time (Woodard, 1959). The highest recorded values of urinary hydroxyproline are also found in this disease, and are due to rapid turnover of bone matrix collagen (Chapter 1). Other biochemical measurements which probably parallel the histological overactivity are the non-protein-bound fraction of hydroxyproline in plasma, and an enzyme measured in the serum, proline iminopeptidase (PIP), which can degrade small collagen-derived fragments (Whiteley *et al.*, 1976).

PAGET'S DISEASE OF BONE
CLINICAL ASPECTS

Incidence

Paget's disease is common in Britain, and the accepted overall incidence of about 3 per cent in people older than 40 years is based on radiological or post mortem evidence. Since the incidence increases rapidly with age, this 3 per cent is the average of up to 10 per cent in subjects in their 80s, and 1 per cent in their 40s. Surveys refer to a selected part of the population who have been x-rayed for some medical reason, or who have come to autopsy. The incidence in the whole population is not known. In the UK with a population of 49 million, of whom about 43 per cent are over 40 years of age, approximately 650 000 persons will have Paget's disease and perhaps 5 per cent of these will have symptoms.

The geographical distribution of Paget's disease is remarkable and inexplicable. Thus it is almost unknown in Scandinavia and Japan, common in Australia, but rarely described in the tropics. It is said to be a disorder of the Anglo-Saxon races and their offspring. In Britain there are also geographical differences; thus the highest incidence appears to be in the North and West and in Lancashire towns (Barker and Gardner, 1974; Barker *et al.*, 1976) but it is not closely related to industry or to Northern latitude.

Although there is no doubt that this disorder increases with age, it is not unknown in the 20s and 30s. This statement does not include the very rare condition of 'juvenile' Paget's disease (Chapter 11). It is impossible to say when Paget's disease begins, and the time of onset of symptoms or of diagnosis must be many years after this. The sex difference is not marked, but males appear to be more frequently affected than females.

Signs and symptoms

These are pain, bony deformity, fracture, deafness, nerve compression, heart failure and sarcoma.

Pain.   This is a common symptom but one which is very difficult to interpret, and there is little agreement about its origin or cause. One (extreme) view is that the bone of Paget's disease is not painful unless there is coexistent sarcoma, or fracture. It is probably correct that pain in the femur in a patient who has Paget's disease of the pelvis or femur on the same side may often be due to joint disease. However, it is difficult to be certain whether this joint disease itself is the result of the Paget's disease rather than a coincidental degenerative arthritis. It has been suggested that there is a specific type of hip joint disorder in Paget's disease (Detenbeck, Sim and Johnson, 1973). Khairi *et al.*, (1973) found that the pain correlated with radiological appearances, and that patients with sclerotic lesions or lesions with a coarse trabecular pattern did not have symptoms. Pain was associated with gross deformity with fissure fractures of the long bones, and often with joint involvement. Nevertheless, Paget's disease may itself produce bone pain and a characteristic example is the localized pain (and tenderness) over the midshaft of a deformed tibia. Such pain is often worse when walking. It may occur without any evidence of microfracture, but this is not always easy to exclude (and in contrast multiple microfractures may be painless). There is no certain explanation for the cause of such bone pain, but it is likely to come from the periosteum where the pain fibres of bone are mainly located. It is possible that the reduction of vascularity produced by such agents as calcitonin which rapidly reduce cellular overactivity also reduces periosteal stretching and thus relieves pain; older therapeutic methods of multiple drilling of bone might work in the same way.

Deformity. This may take many forms, from the midshaft bowing and thickening of the tibia (probably the commonest) to the enlarged 'tam-o-shanter' skull, falling over the ears, and the clavicles like coat-hangers (*Figure 5.2*).

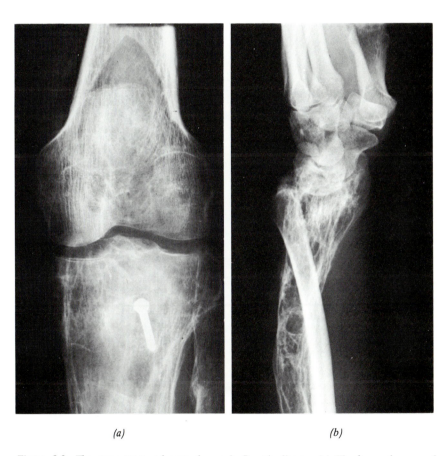

(a)   (b)

Figure 5.2. The appearance of some bones in Paget's disease. (a) The lower femur and upper tibia in a man of 35 years, who first fractured his tibia at the age of 30. The tibia is thickened and disorganized, and the tibial tubercle has been replaced with a screw. Interestingly the disorder appears to have crossed the joint, and the lower femur shows a 'flame-shaped' area of bone resorption. Radiographs of the lower tibia of this patient are shown in Figure 11.4b. (b) Severe deformity due to Paget's disease of the radius, but not of the ulna. Median nerve compression was a feature

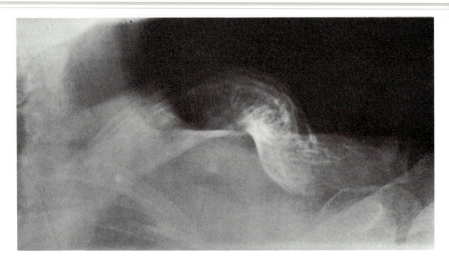

(c)

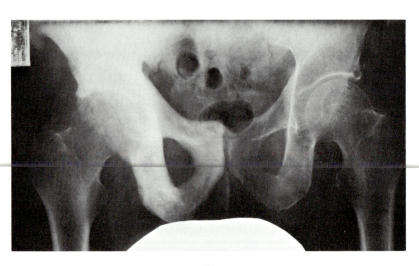

(d)

Figure 5.2 (c) The left clavicle in a man of 50, who had first symptoms of Paget's disease at 38 years. Both clavicles were grossly affected, giving a 'coat-hanger' appearance. (d) The pelvis of a man of 60 years, to show unilateral Paget's disease. Note that the bone is bigger on the affected side, and that the joint lines are not crossed

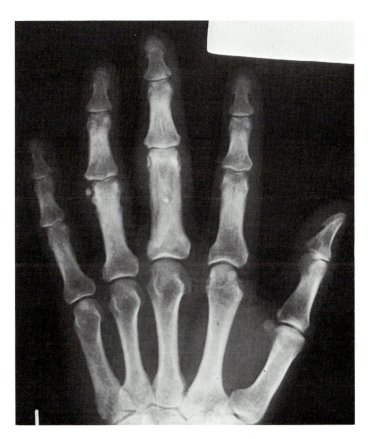

Figure 5.3. A radiograph of the hand of a man of 50 years with Paget's disease who was mistakenly treated for prostatic carcinoma and was given oestrogens for several years. The proximal phalanx on the middle finger is wider (and possibly longer) than normal. The plasma alkaline phosphatase was 41 and acid phosphatase 7.8 K.A. units per 100 ml

Bowing of the long bones and loss of height is common. One should recall that any bone may be affected, including the maxilla (*see* Henneman *et al.,* 1963, Figure 3), the phalanges (*Figure 5.3*), and the ribs.

Nerve compression.   Deafness, often profound, is the commonest nerve damage produced by the bone distortion and overgrowth, which may affect many parts

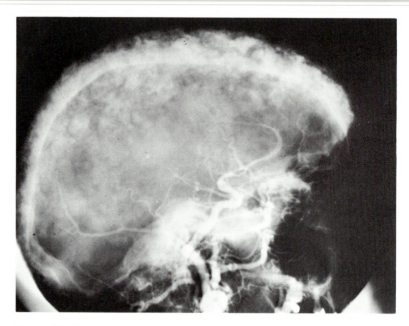

Figure 5.4. The appearance of the skull in a woman of 51 years with diplopia, unsteadiness of gait and occasional difficulty in swallowing due to Paget's disease. She later developed headache, confusion and long-tract signs with evidence of ventricular dilatation. Her symptoms responded only partially to calcitonin but almost completely to craniotomy and a ventricular shunt. A left carotid arteriogram has been done. The skull is thickened and deformed, and the vault has descended towards the base

of the ear. The optic nerve is virtually never to be affected. Diplopia, hysphagia and unsteadiness occurred in a personal case (*Figure 5.4*); and similar symptoms are discussed by Singer (1977). Headache is commonly associated with Paget's disease, but its cause is unknown. Rarely, it may appear to improve with treatment (Hamilton and Quesada, 1973). Spinal cord compression occurs (Sadar, Walton and Gossman, 1972; Cartlidge, McCollum and Ayya, 1972), mainly due to the bony enlargement but increased vascularity and vertebral fracture contribute; sometimes a sharp deformity is present (*Figure 5.5*). Symptoms of claudication of the cauda equina have also been encountered (*Figure 5.6*). In one personal case they disappeared during treatment with EHDP (p. 155). Where there is gross long-bone deformity, peripheral nerves such as the lateral peroneal or median nerve may be distorted (*Figure 5.2b*).

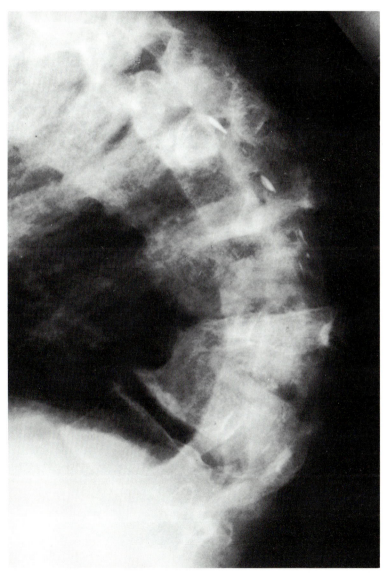

Figure 5.5. A lateral radiograph of the spine in a man of 70 years with spinal cord compression due to Paget's disease. Progression of the symptoms was possibly halted by treatment with EHDP, but a laminectomy was eventually necessary to relieve the compression. A myelogram has been done

141

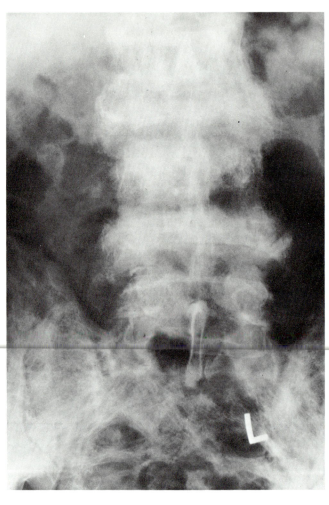

Figure 5.6. The appearance of the lumbar vertebrae in a patient with symptoms of cauda equina claudication probably due to Paget's disease. Only one lumbar vertebra is not grossly abnormal

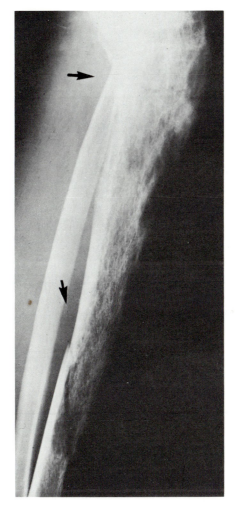

Figure 5.7. Two fractures (arrowed) in a man of 40 years with resorptive Paget's disease in the left tibia. The lower fracture is near to the junction of normal and resorbed bone

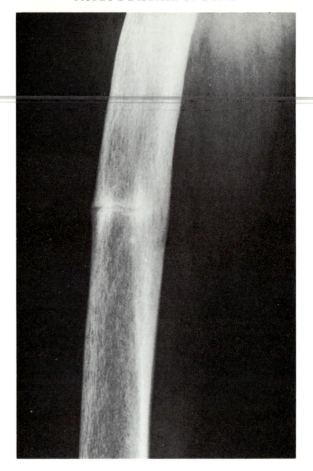

Figure 5.8. Lateral view of partial fracture in a woman of 60 years with probable Paget's disease. The appearance resembles that of a Looser's zone but is on the convexity of the bone, and neither osteomalacia nor hypophosphatasia were present. There was considerable pain until complete fracture occurred and an intramedullary rod was inserted

Fracture. This may occur at any time, either through a resorbing bone, sometimes in a relatively young patient (*Figure 5.7*), or through the thickened but weak bone of an older subject, where a microfracture may be a preceding event. Such a microfracture, on the convexity of a long bone, may be painful until the fracture is complete and the bone subsequently stabilized (*Figure 5.8*).

Heart failure. Cardiac output can be increased in patients with Paget's disease (Howarth, 1953) and rarely heart failure of the high output variety occurs. This is generally considered to be due to the demonstrated increased blood flow (Wootton, Reeve and Veall, 1976) through the affected bones and is presumably

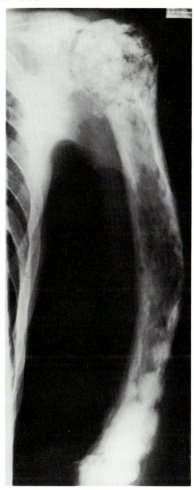

Figure 5.9. The appearance of a sarcoma in the upper end of the left humerus in a man of 70 years known to have Paget's disease for many years, who developed localized pain and swelling. The initial lesion at the proximal end of the humerus was treated with radiotherapy, but secondary deposits have appeared around the lower end. The lung fields remained clear

related to the activity of the bone disease and to the number of bones affected. However, Crosbie, Mohamedally and Woodhouse (1975) could find no correlation between cardiac output and THP excretion. The presence of large shunts with wide vessels has been effectively excluded (Rhodes *et al.*, 1972).

Sarcoma. This is uncommon. The percentage of patients with Paget's disease who develop it is not well established (Barry, 1969) but is probably not more than 1 per cent. It is the main cause of the increase in bone sarcomata in patients over 50 years of age. The presence of a sarcoma is suggested by localized and progressive pain in the bone in a patient with long-standing Paget's disease. Swelling may also occur. Although Paget's disease is commonest in the spine and pelvis, sarcoma associated with it is said to be found proportionately more often in the humerus (Barry, 1969) (*Figure 5.9*).

Associated disorders. Patients with Paget's disease may have many other disorders, since this disease tends to occur in the elderly. Renal stones may be more common than in the rest of the population, and Franck *et al.,* (1974) described an increased incidence of gouty arthritis, hyperuricaemia and rheumatoid arthritis resembling ankylosing spondylitis. Calcific periarthritis and articular chondrocalcinosis also occurred.

Singer (1977) in a description of the non-neoplastic medical complications of Paget's disease, emphasizes the apparently variable incidence of chrondrocalcinosis in this disorder and speculates that this could be due to the variable concentrations of alkaline phosphatase (so that a high alkaline phosphatase would be related to low pyrophosphate concentrations and a low incidence of chrondrocalcinosis). Since Paget's disease is progressively more common with increasing age, it will occur in association with many other conditions whose incidence also increases with age. These include osteoarthritis, vascular calcification, heart failure and gout.

Biochemistry

In most patients the plasma calcium is normal; when it is increased this can usually be attributed to immobilization, particularly if the disorder is active

TABLE 5.II
Relationship between biochemical values and histological measurements in Pagetic bone.
(From Walton *et al.,* 1977b)

Variables	Sex	n	t	P
[Pi] vs osteoid area	M	17	−0.57	<0.05
	F	9	−0.24	NS
	M & F	26	−0.54	<0.01
[Pi] vs osteoid width	M	17	−0.68	<0.01
	F	9	−0.19	NS
	M & F	26	−0.59	<0.01
[Ca] vs osteoid area	M & F	26	−0.05	NS
[Pi] × [Ca] vs osteoid area	M & F	26	−0.036	NS
25(OH)D vs osteoid area	M & F	19	0.05	NS
log-THP vs active resorption surfaces (% total trabecular surfaces)	M & F	19	0.16	NS
log-AP vs active deposition surfaces (% total trabecular surfaces)	M & F	27	0.38	NS

Biochemical values are those in plasma, with the exception of THP (urine total hydroxyproline excretion). AP = alkaline phosphatase.

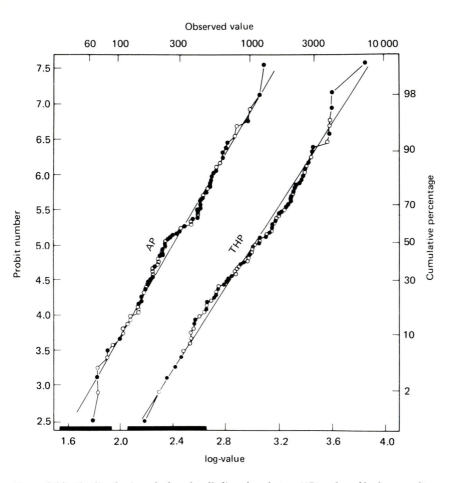

Figure 5.10. The distribution of values for alkaline phosphatase (AP) and total hydroxyproline (THP) in a personal series of patients with Paget's disease. The log values are plotted on probability paper and their linearity demonstrates that this is a homogeneous population. ● male; ○ female. (From Walton, 1977a)

(Russell and Smith, 1973), or to coexistent hyperparathyroidism. The plasma phosphate may be in the upper part of the normal range and according to Bijvoet and de Vries (1974) is proportional to the urinary total hydroxyproline excretion and to the prevailing rate of bone turnover. Walton *et al.,* (1977b) were not able to confirm this, but did find that the plasma phosphate is inversely proportional to the amount of osteoid in Paget's disease (Table 5.II).

The activity of Paget's disease is assessed biochemically by the plasma alkaline phosphatase and the urine total hydroxyproline. These measurements appear to be log-normally distributed (*Figure 5.10*) and are closely related to each other

(Franck *et al.*, 1974; Walton *et al.*, 1977a; *Figure 5.11*). The total hydroxyproline is composed of a dialysable small peptide fraction (about 90 per cent) derived mainly from collagen breakdown, and of large polypeptides which are non-dialysable (Krane, Munoz and Harris, 1970). These polypeptides which may be related to collagen synthesis comprise about 10 per cent of the total urinary hydroxyproline. The non-protein-bound hydroxyproline in the plasma also provides an index of bone activity; it is less easy to measure and less accurate

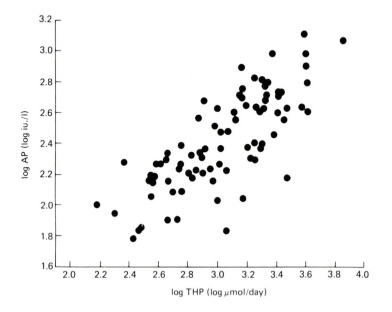

Figure 5.11. To show the close relationship between AP and THP in Paget's disease. The values are log-transformed because the linear values are not normally distributed. (From Walton et al., 1977a)

than the hydroxyproline of urine. The close relationship of alkaline phosphatase (AP) and urine total hydroxyproline (THP) reflects the known (but ill-understood) close relation between bone formation and resorption. It was, however, unexpected to find (Walton *et al.*, 1977b) that in our patients there was no correlation between THP and active resorption, and AP and active formation (Table 5.II) since earlier reports on smaller numbers of patients had suggested otherwise.

Radiology

Paget's disease produces a large number of different radiological appearances and should always be considered where the diagnosis of any bone disorder is in doubt. It seems generally true that the appearance is different early in the disease and in the young person, compared with late in the disease and in the elderly. In the first group resorption is a major feature, either with an advancing front (*Figures 5.2a* and *5.7*) or an area of apparently circumscribed osteoporosis. The shape of the bone may rapidly change, producing, for instance, large bumps on the tibia. In the second group the bone is often dense and deformed with coarse trabeculae; the affected bone is always larger than its fellow and this is a particularly useful sign in the spine which may show different appearances according to the extent of the disease, and in the phalanges (*Figure 5.3*). Isolated bones are often

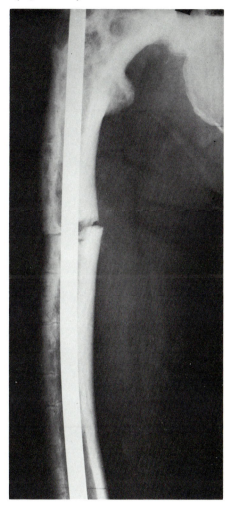

Figure 5.12. Multiple microfractures in the right femur of an 80-year-old man with Paget's disease. Fracture has occurred through one of them and the intramedullary rod has subsequently been inserted

involved; the reason why certain bones may be picked out, and why the condition does not usually spread across joints (*compare, however, Figures 5.2a* and *5.2d*) is quite unknown. It is important to distinguish between micro-fractures on the convexity of the long bones which are often multiple (*Figure 5.12*) and the pseudofractures or Looser's zones of osteomalacia, which occur on the concavity and are usually single. Single fractures on the convexity may be seen without obvious cause as 'stress fractures' and rarely in hypophosphatasia (Chapter 11).

Doyle *et al.,* (1974 a, b) have discussed the radiological features of Paget's disease which may be useful in the assessment of treatment, and the original papers should be consulted. It is clear that we need to know a lot more about the normal development of the radiological findings. Thus the healing of osteoporosis circumscripta of the skull demonstrated in Figure 2 of Doyle *et al.,* (1974a) and Figure 1 of de Rose *et al.,* (1974) is undoubted, but the former authors consider that this might occur normally, whereas the latter describe it as 'dramatic improvement' (presumably due to treatment). The radiological appearance of a sarcoma developing in an area of Paget's disease (*Figure 5.9*) is said to be mainly destructive or lytic. Barry (1969) points out that since most tumours appear to arise in the medulla, the radiological features of sarcoma in an earlier age group, such as 'sunray' spicules of new bone, and Codman's triangle, are not often seen.

Other investigations

Bone biopsy

The diagnosis of Paget's disease is usually obvious but when doubt exists iliac bone biopsy is necessary. Since Paget's disease is a patchy disease the bone obtained may be normal, but such a biopsy will exclude important alternative generalized bone diseases such as osteomalacia (*see* Chapter 4).

Bone scan

Areas of active Paget's disease will tend to selectively concentrate bone scanning agents (Chapter 2), such as technetium-labelled diphosphonate (Figure 53 in Singer, 1977).

External balance and isotopes

The overall calcium balance can be negative, positive or zero and this may reflect the phase of the bone disease (or the technical competence of the investigators). Isotope studies demonstrate the expected rapid bone turnover.

Other biochemical measurements

Further biochemical measurements are being developed which may be of use in following the effect of treatment. These include proline iminopeptidase (PIP) (*Figure 5.13*), an enzyme in plasma which may break down smaller collagen fragments, which is increased in active Paget's disease and is unaltered by dietary collagen (Whiteley *et al.,* 1976); and a circulating α_2HS glycoprotein (Chapter 1) which appears to be incorporated into newly mineralized bone. The concentration of this glycoprotein in plasma tends to increase with treatment, when the excessive formation is suppressed, and the change with time is therefore opposite to that of alkaline phosphatase.

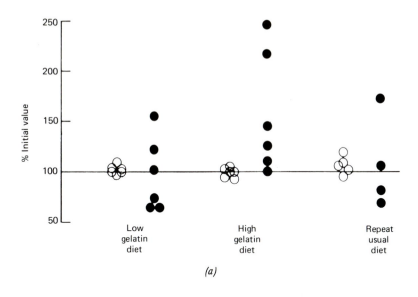

(a)

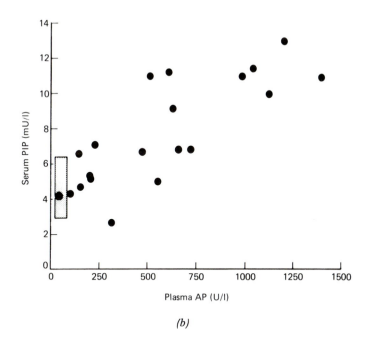

(b)

Figure 5.13. To demonstrate that plasma proline iminopeptidase (PIP) is: (a) independent of dietary gelatin; (b) correlated with plasma AP in Paget's disease. A similar relationship exists for THP (Whiteley et al., 1976). In (a), open circles refer to repeated estimations of PIP and closed circles to repeated estimations of urinary hydroxyproline. The measurements are on the same subjects on different diets. In (b) the rectangular area outlines the normal range

DIFFERENTIAL DIAGNOSIS

In the usual case the combination of the physical signs (often with a deformed warm bone), the increase in alkaline phosphatase with otherwise normal routine biochemistry, and the characteristic x-ray appearance should leave no doubt about the diagnosis. However, Paget's disease can be confused with:

(a) Other conditions with a raised alkaline phosphatase; such as osteomalacia (Chapter 4), parathyroid bone disease(Chapter 6), and, rarely, idiopathic hyperphosphatasia (Chapter 11). The distinction from osteomalacia should be easy to make from the biochemical differences and, if necessary, can be settled outright by bone histology. Similarly, the diagnosis of hyperparathyroidism with bone disease should present no difficulty. However both may coexist with Paget's disease. Whether idiopathic hyperphosphatasia is a form of Paget's disease occurring in childhood or is a different disease, is not clear. However, often there are affected siblings, suggesting recessive inheritance, and the bones are grossly deformed from childhood.

(b) Conditions with a similar radiological appearance; Paget's disease may simulate most other bone diseases, but one should particularly note the differential diagnosis from metastatic bone disease, fibrous dysplasia (Chapter 11) and, occasionally, chronic osteomyelitis. Osteoblastic secondaries from the prostate may suggest Paget's disease but the bones are not enlarged or deformed and the plasma acid phosphatase is increased. In Paget's disease the acid phosphatase may be increased but far less than the alkaline phosphatase. Rarely, osteolytic deposits may be difficult to distinguish from Paget's disease in which a sarcoma has developed. In the generalized polyostotic form of fibrous dysplasia (Chapter 11) the alkaline phosphatase and urinary hydroxyproline may be slightly increased, but the unilateral bone lesions and pigmentation, the sexual precocity in females and occasional thyrotoxicosis will distinguish this disorder. In the localized form the alkaline phosphatase is said to be normal but real difficulties in distinction from early Paget's disease may occur (Chapter 11, *Figure 11.4*) and bone biopsy may be required.

TREATMENT

The majority of patients with Paget's disease have no complaints and require no treatment. However, it is a common disorder and there must be many thousands in the UK with symptoms. Probably the commonest symptom which requires treatment is pain.

The treatments available for Paget's disease include, first, those which reduce the symptoms and ameliorate the complications, and, second, those which in addition appear to suppress the excessive bone turnover, such as mithramycin, glucagon, calcitonin and diphosphonates.

Of the former, the pain of Paget's disease often responds better to simple analgesics than to anti-inflammatory drugs, and in the past various empirical procedures such as drilling, curettage, osteotomy or radiotherapy, have had some success. For severe deformity of long bones osteotomies, with or without intramedullary nailing, may be considered, and spinal cord involvement will require decompression. Singer (1977) discusses the usefulness of surgery, including ventriculojugular shunts where there is evidence of hydrocephalus (*see Figure 5.4*). Whatever surgical procedure is adopted, the period of immobility must be as short as possible, to avoid excessive bone loss, hypercalcaemia and hypercalcuria.

In the latter, calcitonin and diphosphonates have attracted most interest, because of the possibility that they might prevent long-term complications as well as control symptoms.

Mithramycin is an antimitotic drug which prevents RNA synthesis and which, together with actinomycin (Fennelly and Groarke, 1971), has been used in the treatment of malignant disease. When this is given to patients with Paget's disease there is biochemical evidence of a rapid fall in cellular activity together with improvement in pain (Condon *et al.*, 1971). The effect may be only temporary and repeated doses are necessary. Since it has to be given by intravenous infusion and is potentially toxic it has not been widely used. However Hadjipavlou and his colleagues (1977) have recently described the striking effects of mithramycin on pain in patients with Paget's disease. In patients who had not had previous treatment they found that mithramycin, in a dose of 15 mg/kg body weight, alleviated all types of pain, including severe headache and back pain. Pain relief lasted longer than biochemical suppression. In addition to pain relief there was also evidence of improvement in symptoms attributable to spinal stenosis. Mithramycin was found to cause transient nausea and diarrhoea, which could be controlled by antihistamines, anti-emetics or opiates, and not more than 3 courses of treatment were given, because of possible side-effects. This study combined mithramycin treatment with calcitonin and with glucagon, concluding that mithramycin with calcitonin was the most effective combination.

Glucagon reduces the raised alkaline phosphatase and hydroxyproline in Pagetic patients (Condon, 1971). It is probably effective because it stimulates calcitonin production, but it also has to be given intravenously and is not always well tolerated.

It is impossible to give a firm statement of the therapeutic position of calcitonin and diphosphonates at the moment; since diphosphonates have until recently not been generally available, whereas most calcitonins have, they are of less general interest. It may be that the most effective treatment will be a combination of these two types of drugs. Descriptions of the effects of diphosphonates are given by Altman and his colleagues (1973), and Russell and others (1974), of calcitonin in the *British Medical Journal* (1975), and of combined treatment by Hosking *et al.*, (1976).

Table 5.III lists some of the articles on the medical treatment of Paget's disease together with the main conclusions and demonstrates the considerable work on calcitonin. The reader will find claims of 'striking' and 'dramatic' improvement in some studies and not others. On the whole these reflect the interpretation placed on the results by the authors. The known high placebo response is not always taken into account, and is likely to be more marked for something that is injected rather than taken orally.

The various calcitonins owe their effect in Paget's disease to the rapid reduction in osteoclastic bone resorption which they produce. The most convincing effects have been produced in patients given human calcitonin, where the plasma alkaline phosphatase and urinary hydroxyproline can be reduced to normal levels (Krane *et al.*, 1973), the bone histology may return towards

TABLE 5.III
Some recorded effects of calcitonin and diphosphonate in Paget's disease

Date	Author	Calcitonin (CT) Type of CT	Effect
1968	Bijvoet, Van der Sluys Veer and Jansen	Porcine	Plasma Ca, urine THP reduced by CT i.v. Effect depends on rate of bone resorption
1971	Shai, Baker and Wallach	Porcine	Variable. Claimed neurological improvement
1970	Haddad, Birge and Avioli	Porcine	Unimpressive fall in THP. Temporary increase in proportion of non-dialysable hydroxyproline
1971	Woodhouse et al.	Human	Bone pain, AP, THP, local temperature reduced
1973	Dube et al.	Porcine	Calcitonin antibodies
1973	Walton and Strong	Salmon	No effect on coexistent sarcoma
1973	Krane et al.	Porcine & Salmon	Fall in THP, oligopeptides more than polypeptides
1974a	Doyle et al.	Human	X-ray evidence of dose-related respons
1974b	Doyle et al.	Human	Healing of bones in 'Juvenile Paget's'
1974	Kanis et al.	Salmon	Low dose 3 × weekly. Prolonged pain remission
1974	Moffat, Morrow and Simpson	Porcine	One case, improved hearing
1974	Rojanasathit, Rosenberg and Haddad	Human	Effective in patients resistant to salmon CT. Also biochemical suppression with EHDP
1975	Avramides et al.	Salmon	Once weekly 50 MRC units. Slight effect
1974	de Rose et al.	Porcine & Salmon	Salmon preferable to porcine. 'Dramatic' claims
1975	British Medical Journal		Editorial
1975	Kanis, Fitzpatrick and Strong	Porcine	Effectiveness of repeated short courses
1975	Crosbie, Mohamedally and Woodhouse	Salmon	12 patients. No fall in cardiac output. THP fell
1976	Melick, Ebeling and Hjorth	Porcine	Reported improvement in paraplegia

Date	Author	Diphosphonate (EHDP)	Effect
1973	Altman et al.		47 patients. Dose-related pain improvement
1973	Fleisch and Bonjour		Editorial, New England Journal of Medicine
1974	Russell et al.		Multi-dose study, 47 patients. Histological and biochemical evidence of suppression of activity

Date	Author	CT plus EHDP Type of CT	Effect
1976	Hosking et al.	Human	Effective suppression of disease, without complications of either agent alone

This Table is not intended to be comprehensive, but to indicate the development of the treatment of Paget's disease with CT and EHDP. The current position is reviewed in *Lancet* (1978). *See also* Chapter 12

normal and pain improves (Woodhouse *et al.,* 1971). With the human hormone antibodies to human calcitonin do not develop, but resistance may, and the increased biochemical activity soon returns when calcitonin is stopped. Radio-logical improvement has been most convincing in a young patient with 'juvenile Paget's disease' (Doyle *et al.,* 1974b), in whom remodelling can presumably more easily occur than in adults. Radiological improvement may also appear to occur in adults; the authors (Doyle *et al.,* 1974a) realize the extreme difficulty of assessing this, as the natural history of Paget's disease is so poorly documented.

Other calcitonins are not so effective. With porcine calcitonin response is variable, biochemical escape may occur despite increasing dose, and antibodies (of questionable significance) may develop. Salmon calcitonin is better and Paget's disease may appear to be controlled on small weekly doses. All forms of calcitonin are expensive and have to be given by injection; the injections are often poorly tolerated, and may produce paraesthesiae, nausea, vomiting and sometimes diarrhoea. Some patients prefer to have their calcitonin injection in the evening, preceded by an anti-emetic, so that they may sleep through the side-effects. There is no double blind trial of the effects of calcitonin on symptoms and hence no measurement of the placebo response to non-calcitonin injections.

Diphosphonates are related to pyrophosphate but have a P-C-P backbone rather than the P-O-P one of pyrophosphate. This makes them resistant to the action of naturally occurring phosphatases and pyrophosphatases, and effective orally (Russell, 1975). They have been shown (Fleisch, Russell and Francis, 1969) in animals to reduce bone resorption and turnover and to reduce ectopic calcification. In patients with Paget's disease they produce a dose-related suppression of the biochemical indices of excessive bone activity. This suppression may continue for many months after treatment is stopped. There appears to be a significant reduction in pain when this can be directly attributed to Paget's disease. Since EHDP* (which is the diphosphonate used) can interfere with mineralization when given in sufficient dose, one of the main problems is to find a dose of EHDP which effectively suppresses bone resorption without producing a disorder of mineralization.

Comparing calcitonin and EHDP, there is no convincing evidence that either of these substances reverses spinal cord compression, although isolated reports do suggest an improvement (Shai, Baker and Wallach, 1971; Melick, Ebeling and Hjorth, 1976). There is also no evidence that either substance alters the course of Pagetic sarcoma (Walton and Strong, 1973; Russell and Smith, 1973). Those aspects of the disease which are directly related to the locally excessive bone turnover and increased blood flow such as local warmth over the affected bone and increased uptake of bone-scanning agents may be reduced (Guncaga *et al.,* 1974). The evidence from one group on cardiac output is conflicting (Woodhouse, Crosbie and Mohamedally, 1975; Crosbie, Mohamedally and Woodhouse, 1975). The recorded effects of calcitonin treatment on the deafness of Paget's disease have been variable. However Solomon *et al.,* (1977) found that in a group of patients with Paget's disease of the skull and deafness, those who received continuous calcitonin treatment showed less deterioration of hearing over 9 to 18 months, measured by pure-tone audiometry, than those who did not. No treated patient showed substantial improvement.

*Ethane-1-hydroxy-1, 1-diphosphonate.

Since there is little doubt, especially from the biochemical evidence, that both calcitonin and EHDP, separately or together (Hosking *et al.,* 1976), can reduce the activity of Paget's disease, it is necessary to consider whom to treat. These should include patients with symptoms (particularly pain), young patients (in the hope of preventing progress of the disease) and those with complications (for the same reason).

So far the stated indications for treatment depend largely on personal experience. Thus Singer (1977) suggests that in addition to pain and neurological complications, calcitonin may be used before surgery on Pagetic bone and for metabolic complications (renal calculi, hypercalcaemia, gout, etc). If future work confirms the usefulness of various phosphonates, the indications for an oral treatment are likely to be wider than for one given by injection.

It is appropriate to remember that other forms of treatment can also reduce the biochemical measures of bony overactivity — these include steroids, aspirin and fluoride. Fluoride stimulates new bone formation and is not therefore correct treatment where there is nerve compression, but it might reasonably be considered where there is predominantly resorptive disease. Improvement in bone pain and calcium balance has been reported (Purves, 1962). However Nagant de Deuxchaisnes and Krane (1964) found no convincing effects on alkaline phosphatase or urinary hydroxyprline. At present there is not sufficient published evidence to recommend its use, bearing in mind the potential hazards of prolonged high fluoride intake.

More recently the efficacy of calcium infusion or a high calcium, low phosphate diet, has been shown (Sekel, 1973; Evans 1977) and symptoms have also been considerably improved. It may be that the effect is directly due to the increase in plasma calcium or due to the stimulation of calcitonin production.

A recent editorial in the *Lancet* (1978) usefully summarizes the medical treatment of Paget's disease since Bijvoet, Van der Sluys Veer and Jansen, (1968) first described the effect of calcitonin in this disorder. The abnormal mineralization recorded in the early work on EHDP is unlikely to be a future problem with lower doses or combined treatment, and other diphosphonates currently being studied do not have this effect.

DISCUSSION AND FUTURE PROBLEMS

The individual with generalized Paget's disease may be severely disabled, and the disorder must have been coming on for many years. The problems of such a patient are easy to state but difficult to solve. We must enquire:— What is this disease?; How has it progressed?; How active is it?; Can it be controlled?; and importantly; Would it have been an advantage to treat it years ago?

The cause of Paget's disease eludes us. The striking geographical and sometimes familial incidence suggests that both the environment and inheritance are important. It may be that the connective tissue is abnormal (McKusick, 1972). The polymeric collagen of the skin differs from age-matched controls (Francis and Smith, 1974), but the evidence is unconvincing and the observation may be secondary to the unidentified primary disorder. Plasma calcitonin is normal (Kanis, Heynen and Walton, 1977). Paget's disease behaves in some respects like a multicentric neoplasm. There are also recent reports (Rebel *et al.,* 1977; Mills and Singer, 1976) of organized inclusions in the osteoclasts of Pagetic bone which have been thought to indicate a virus. It is not clear how specific these findings are.

We are very ignorant of the natural history of Paget's disease. Particularly there seems a gap between localized and generalized Paget's disease; patients may be followed for years with, for instance, a single tibia affected without evidence of spread to other bones; other patients may be seen for the first time with generalized disease and no reliable history of previous local deformity. As in osteoporosis (and many other conditions) longitudinal studies to settle this point are most important. This is particularly so for individual prognosis; for instance, many patients with localized Paget's disease of the tibia want to know if the bones of their skull, and by inference its contents, are going to be affected.

At least we can now assess its activity more accurately; measurements of hydroxyproline excretion, which correlate so well with the alkaline phosphatase, can tell us that excessive bone turnover is occurring, but cannot readily differentiate the early resorptive phase from the later phase of excessive bone formation.

It is becoming increasingly obvious that this disease can be controlled, although the evidence for this is mainly biochemical, and repeated treatment may be needed. It is very important to establish whether complications may be prevented, but this will require much longer studies than now exist. In this respect the expectations have probably been too great. For instance it is naive to think that medical treatment could have any but the slightest effect on cord compression produced by vertebrae which have been enlarging and deforming for years. Certainly there is more hope in the young, where bone turnover and remodelling is normally more rapid than in the adult.

Finally, there is much to be done in the practical treatment of Paget's disease. Combined treatment, not only between calcitonin and phosphonate, but as an adjunct to surgery (for instance to reduce vascularity before hip replacement), may be acceptable and useful in the future.

SUMMARY

Paget's disease of bone has emerged from the descriptive era. Biochemical advance has enabled scientific assessment of new treatment which seems capable of suppressing the disease. Search for its cause, and prevention of its complications, are now of first importance.

REFERENCES

Altman, R. D., Johnston, C. C., Khairi, M. R. A., Wellman, H., Serafini, A. N. and Sankey, R. R. (1973). Influence of disodium etidronate on clinical and laboratory manifestations of Paget's disease of bone (osteitis deformans). *New Engl. J. Med.* **289**, 1397–1384

Avramides, A., Flores, A., DeRose, J. and Wallach, S. (1975). Treatment of Paget's disease of bone with once-a-week injections of salmon calcitonin. *Br. med. J.* **3**, 632

Barker, D. J. P. and Gardner, M. J. (1974). Distribution of Paget's disease in England, Wales and Scotland and a possible relationship with vitamin D deficiency in childhood. *Br. J. prev. soc. Med.* **28**, 226–232

Barker, D. J. P., Clough, P. W. L., Guyer, P. B. and Gardner, M. J. (1977). Paget's disease of bone in fourteen British towns. *Br. med. J.* **1**, 1181–1183

Barry, H. C. (1969). *Paget's Disease of Bone*. Edinburgh and London: E. & S. Livingstone Ltd

Bijvoet, O. L. M. and de Vries, H. R. (1974). Plasma phosphate in Paget's disease. *Lancet* **1**, 1283–1284

REFERENCES

Bijvoet, O. L. M., Van der Sluys Veer, J. and Jansen, A. P. (1968). Effects of calcitonin on patients with Paget's disease, thyrotoxicosis or hypercalcaemia. *Lancet* **1**, 876–881

British Medical Journal (1975). Paget's disease and calcitonin. **3**, 505–506

Cartlidge, N. E. F., McCollum, J. P. K. and Ayyar, R. D. A. (1972). Spinal cord compression in Paget's disease. *J. Neurol. Neurosurg. Psychiat.* **35**, 825–828

Condon, J. R. (1971). Glucagon in the treatment of Paget's disease of bone. *Br. med. J.* **1**, 719–721

Condon, J. R., Reith, S. B. M., Nassim, J. R., Millard, F. J. C., Hilb, A. and Stainthorpe, E. M. (1971). Treatment of Paget's disease of bone with mithramycin. *Br. med. J.* **1**, 421–423

Crosbie, W. A., Mohamedally, S. M. and Woodhouse, N. J. Y. (1975). Effect of salmon calcitonin on cardiac output oxygen transport and bone turnover in patients with Paget's disease. *Clin. Sci. molec. Med.* **48**, 537–540

De Rose, J., Singer, F. R., Avramides, A., Flores, A., Dziadiw, R., Baker, R. K. and Wallach, S. (1974). Response of Paget's disease to porcine and salmon calcitonin. Effects of long-term treatment. *Am. J. Med.* **56**, 858–866

Detenbeck, L. C., Sim, F. H. and Johnson, E. W. (1973). Symptomatic Paget disease of the hip. *J. Am. med. Ass.* **224**, 213–217

Doyle, F. H., Pennock, J., Greenberg, P. B., Joplin, G. F. and MacIntyre, I. (1974a) Radio-logical evidence of a dose-related response to long-term treatment of Paget's disease with human calcitonin. *Br. J. Radiol.* **47**, 1–8

Doyle, F. H., Woodhouse, N. Y. J., Glen, A. C. A., Joplin, G. F. and MacIntyre, I. (1974b). Healing of bones in juvenile Paget's disease treated by human calcitonin. *Br. J. Radiol.* **47**, 9–15

Dube, W. J., Goldsmith, R. S., Arnaud, S. B. and Arnaud, C. (1973). Development of antibodies to porcine calcitonin during treatment of Paget's disease of bone. *Proc. Mayo Clin.* **48**, 43–46

Evans, R. A. (1977). A cheap oral therapy for Paget's disease of bone. Proc. 12th Europ. Symp. Calcif. Tissues. *Calcif. Tissue Res.* (Suppl.) **22**, 287–291

Fennelly, J. J. and Groarke, J. F. (1971). Effect of actinomycin D on Paget's disease of bone. *Br. med. J.* **1**, 423–426

Fleisch, H. and Bonjour, J. P. (1973). Diphosphonate treatment in bone disease. *New Engl. J. Med.* **289**, 1419–1420

Fleisch, H., Russell, R. G. G. and Francis, M. D. (1969). Diphosphonates inhibit hydroxy-apatite dissolution *in vitro* and bone resorption in tissue culture and *in vivo*. *Science N. Y.* **165**, 1261–1264

Francis, M. J. O. and Smith, R. (1974). Evidence of a generalized connective tissue defect in Paget's disease of bone. *Lancet* **1**, 841–842

Franck, W. A., Bress, N. M., Singer, F. R. and Krane, S. M. (1974). Rheumatic manifestations of Paget's disease of bone. *Am. J. Med.* **56**, 592–603

Guncaga, J., Lauffenburger, Th, Lentner, C., Dambacher, M. A., Haas, H. G., Fleisch, H. and Olah, A. J. (1974). Diphosphonate treatment of Paget's disease of bone. A correlated metabolic calcium kinetic and morphometric study. *Hormone Metabolic Res.* **6**, 62–69

Haddad, J. G., Birge, S. J. and Avioli, L. V. (1970). Effects of prolonged thyrocalcitonin administration on Paget's disease of bone. *New Engl. J. Med.* **283**, 549–555

Hadjipavlou, A. G., Tsoukas, G. M., Siller, T. N., Danais, S. and Greenwood, F. (1977). Combination drug therapy in Paget's disease of bone. *J. Bone Jt Surg.* **59A**, 1045–1051

Hamilton, C. R. and Quesada, O. (1973). Paget's disease of the skull and migraine headache. *Johns Hopkins med. J.* **132**, 179–185

Henneman, P. H., Dull, T. A., Avioli, L. V., Bastomsky, C. H. and Lynch, T. N. (1963–64). Effects, of aspirin and corticosteroids on Paget's disease of bone. *Trans. Stud. Coll. Physns. Philad.* **31**, 10–25

Hosking, D. J., Bijvoet, O. L. M., Aken, J. V. and Will, E. J. (1976). Paget's bone disease treated with diphosphonate and calcitonin. *Lancet* **1**, 615–617

Howarth, S. (1953). Cardiac output in osteitis deformans. *Clin. Sci.* **12**, 271–275

Kanis, J. A., Fitzpatrick, K. and Strong, J. A. (1975). Treatment of Paget's disease of bone with porcine calcitonin: clinical and metabolic responses. *Q. Jl. Med.* **44**, 399–413

Kanis, J. A., Heynen, G. and Walton, R. J. (1977). Plasma calcitonin in Paget's disease of bone. *Clin. Sci. molec. Med.* **52**, 329–332

Kanis, J. A., Horn, D. B., Scott, R. D. M. and Strong, J. A. (1974). Treatment of Paget's disease of bone with synthetic salmon calcitonin. *Br. med. J.* **3**, 727–731

REFERENCES

Khairi, M. R. A., Wellman, H. N., Robb, J. A. and Johnston, C. C. Jr. (1973). Paget's disease of bone (Osteitis deformans): symptomatic lesions and bone scan. *Ann. intern. Med.* **79**, 348–351

Krane, S. M., Munoz, A. J. and Harris, E. D. (1970). Urinary polypeptides related to collagen synthesis. *J. Clin. Invest.* **49**, 716–729

Krane, S. M., Harris, E. D., Singer, F. R., and Potts, J. T. (1973). Acute effects of calcitonin on bone formation in man. *Metabolism* **22**, 51–58

Lancet (1978). Ten Years Treatment for Paget's Disease (Editorial). **1**, 914–915

McKusick, V. A. (1972). *Heritable Disorders of Connective Tissue.* St. Louis: C. V. Mosby Co

Melick, R. A., Ebeling, P. and Hjorth, R. J. (1976). Improvement in paraplegia in vertebral Paget's disease treated with calcitonin. *Br. med. J.* **1**, 627–628

Mills, B. G. and Singer, F. R. (1976). Nuclear inclusions in Paget's disease of bone. *Science, N. Y.* **194**, 201–202

Moffatt, W. H., Morrow, J. D. and Simpson, N. (1974). Effect of calcitonin therapy on deafness associated with Paget's disease of bone. *Br. med. J.* **4**, 203

Nagant de Deuxchaisnes, C. and Krane, S. M. (1964). Paget's disease of bone: clinical and metabolic observations. *Medicine* **43**, 233–266

Purves, M. J. (1962). Some effects of administering sodium fluoride to patients with Paget's disease. *Lancet* **2**, 1188

Rebel, S., Malkani, K., Basle, M. and Bregeon, Ch. (1977). Is Paget's disease of bone a viral infection? Proc. 12th Europ. Symp. Calcif. Tissues. *Calcif. Tissue Res.* (Suppl.) **22**, 283–286

Rhodes, B. A., Greyson, N. D., Hamilton, C. R., White, R. I., Giargiana, F. A., Wagner, H. N. (1972). Absence of anatomic arteriovenous shunts in Paget's disease of bone. *New Engl. J. Med.* **287**, 686–689

Rojanasathit, S., Rosenberg, E. and Haddad, J. G. (1974). Paget's bone disease: response to human calcitonin in patients resistant to salmon calcitonin. *Lancet* **2**, 1414–1415

Russell, R. G. G. (1975). Diphosphonates and polyphosphates in Medicine. *Br. J. Hosp. Med.* **14**, 297–314

Russell, R. G. G. and Smith, R. (1973). Diphosphonates. Experimental and clinical aspects. *J. Bone Jt Surg.* **55B**, 66–68

Russell, R. G. G., Smith, R., Preston, C., Walton, R. J. and Woods, C. G. (1974). Diphosphonates in Paget's disease. *Lancet* **1**, 894–898

Sadar, E. S., Walton, R. J. and Gossman, H. H. (1972). Neurological dysfunction in Paget's disease of the vertebral column. *J. Neurosurg.* **37**, 661–665

Sekel, R. (1973). Calcium infusion in painful Paget's disease of bone. *Lancet* **1**, 372–373

Shai, F., Baker, R. K. and Wallach, S. (1971). The clinical and metabolic effects of porcine calcitonin on Paget's disease of bone *J. clin. Invest.* **50**, 1927–1940

Singer, F. R. (1977). *Paget's Disease of Bone.* New York and London: Plenum Medical Book Co

Solomon, L. R., Evanson, J. M., Canty, D. P. and Gill, N. W. (1977). Effect of calcitonin treatment on deafness due to Paget's disease of bone. *Br. med. J.* **2**, 485–487

Walton, I. G. and Strong, J. A. (1973). Calcitonin and osteogenic sarcoma. *Lancet* **1**, 887–888

Walton, R. J., Preston, C. J., Bartlett, M., Smith, R. and Russell, R. G. G. (1977a). Biochemical measurements in Paget's disease of bone. *Europ. J. Clin. Invest.* **1**, 37–39

Walton, R. J., Woods, C. G., Russell, R. G. G., Kanis, J. A. and Clark, M. B. (1977b). Histological measurements in Paget's disease of bone. Proc. 12th Europ. Symp. Calcif. Tissues. *Calcif. Tissue Res.* (Suppl.) **22**, 295–297

Whiteley, J., Francis, M. J. O., Walton, R. J. and Smith, R. (1976). Serum proline imino peptidase activity in normal subjects and in patients with Paget's disease of bone. *Clin. chim. Acta* **71**, 157–163

Woodard, H. Q. (1959). Long-term studies of the blood chemistry in Paget's disease of bone. *Cancer* **12**, 1226–1237

Woodhouse, N. J. Y., Bordier, P., Fisher, M., Joplin, G. F., Reiner, M., Kalu, D. N., Foster, G. V. and MacIntyre, I. (1971). Human calcitonin in the treatment of Paget's bone disease.

Woodhouse, N. J. Y., Bordier, P., Fisher, M., Joplin, G. F., Reiner, M., Kalu, D. N., Foster, G. V. and MacIntyre, I. (1971). Human calcitonin in the treatment of Paget's bone disease. *Lancet* **1**, 1139–1143

Wootton, R., Reeve, J. and Veall, N. (1976). The clinical measurement of skeletal blood flow. *Clin. Sci. molec. Med.* **50**, 261–268

6

Parathyroids and Bone Disease

INTRODUCTION

Among the standard works on metabolic bone disease, those dealing with parathyroid overactivity have a prominent place, justified more by its physiological importance than by its frequency. When primary hyperparathyroidism was first identified, patients were most often diagnosed because of their bone disease (for history *see* Fourman and Royer (1968) and Goldman, Gordan and Roof (1971)). It was subsequently realized that patients with primary hyperparathyroidism might present in other ways, for instance, with symptoms due to renal stones, pancreatitis or hypercalcaemia, and that only a minority (probably less than 20 per cent) had significant clinical bone disease. Thus in comparison to Paget's disease or osteoporosis, bone disease due to primary hyperparathyroidism is rare. There have, however, been many clinical and physiological advances in our understanding of parathyroid hormone and its effects which are relevant to the skeleton.

The physiology can conveniently be dealt with in one section, but the clinical aspects of hyper- and hypoparathyroidism require separate consideration. A recent useful account is that of Tomlinson and O'Riordan (1978) and further details may be found in the writings of Fourman and Royer (1968); Potts and Deftos (1974); Goldman, Gordan and Roof (1971); Robinson (1974); Paterson (1974); Vaughan (1975); and *Clinics in Endocrinology and Metabolism* (1974). For those who have time the critical four-part review by Parfitt (1976), referred to in Chapter 1, is outstanding. Earlier work such as that of Hunter and Turnbull (1931) deals particularly with parathyroid bone disease. This chapter does not provide a comprehensive account of parathyroid disorders, but summarizes the many conditions which may arise from a disturbance in its function.

PHYSIOLOGY

Parathyroid hormone (PTH) is an 84 amino acid polypeptide whose secretion by the parathyroid glands is controlled by the circulating level of ionized calcium (Aurbach and Potts, 1967; Arnaud, 1973; and *see* Chapter 1). A reduction in the plasma-ionized calcium stimulates PTH secretion which by its effect on its target organs tends to restore the plasma calcium to normal, thus removing the stimulation.

PTH itself is secreted into the plasma as the whole molecule but circulates as fragments. The main fragments are the C-terminal and N-terminal fragments, named according to the end of the PTH molecule from which they are derived. The C-terminal portion of the molecule appears to have a long half-life but to be biologically relatively inert; the N-terminal portions have opposite properties.

The known target organs of PTH or its fragments are the renal tubules, the bones and the intestine. The main effects of PTH are to lower the renal phosphate 'threshold' (i.e. to reduce the maximum tubular reabsorption of phosphate, the TmP, in relation to glomerular filtration rate), to stimulate osteoclastic bone resorption, and to stimulate renal reabsorption of calcium and calcium absorption across the small intestine.

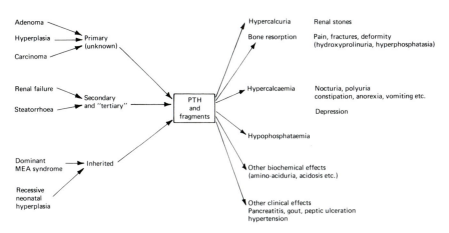

Figure 6.1. To demonstrate some causes and effects of parathyroid overactivity

The effects of PTH on the renal tubule and on the osteoclast appear to be mediated by the membrane-bound adenyl cyclase system which stimulates the conversion of ATP to cyclic AMP (Butcher, 1968). After the injection of exogenous PTH, plasma and urine cAMP increase. PTH appears also to increase the conversion of 25(OH)D to 1,25(OH)$_2$D which may be responsible for its stimulation of calcium transport across the gut. Under certain circumstances, PTH may temporarily stimulate osteoblastic activity, and it has been claimed that certain synthetic fragments increase bone mass in osteoporosis (p. 84).

Because of the presence of PTH fragments in the blood and the different systems used for PTH immunoassay, there has been much controversy about their usefulness. There is also dispute about the amino-acid sequence of part of the molecule. Despite these uncertainties, the diagnosis of hyperparathyroidism is becoming more accurate and separate hypoparathyroid disorders have also been identified, due to proved or suspected blocks in the metabolism of PTH.

Overactivity of the parathyroids (*Figure 6.1*) may be primary (or appear to be so), most often due to adenoma or hyperplasia (p. 163), secondary, due to prolonged hypocalcaemia; and (probably) tertiary, with the formation of an autonomous adenoma after prolonged secondary hyperparathyroidism. Overactivity may be familial, and often part of the multiple endocrine adenoma (MEA) syndrome. Carcinoma is rare.

Underactivity of the parathyroids (*Figure 6.2*) may be due to their removal (surgical hypoparathyroidism) or may be idiopathic, either as an isolated finding or as a familial disorder often associated with evidence of auto-immune disease.

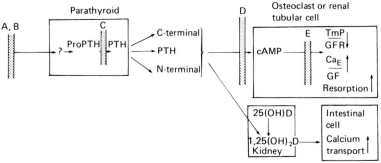

A Surgical hypoparathyroidism.

B Idiopathic hypoparathyroidism.

C Pseudo-idiopathic hypoparathyroidism.

D Pseudohypoparathyroidism Type I.

E Pseudohypoparathyroidism Type II.

Figure 6.2. To demonstrate some causes of parathyroid insufficiency against the background of the normal physiology of parathyroid hormone. For the sake of clarity the effects on the renal hydroxylation of 25(OH)D are separated from other renal effects of PTH. Stippled bars represent lack of formation or postulated metabolic block

A condition has been described (Nusynowitz and Klein, 1973) in which there is evidence of hypoparathyroidism despite high circulating levels of immunologically reacting PTH, possibly due to failure of conversion of proparathyroid hormone (proPTH) to parathyroid hormone (pseudo-idiopathic hypoparathyroidism). There are also syndromes with skeletal abnormalities, hypocalcaemia and hyperphosphataemia where the cAMP response to exogenous PTH is absent (pseudohypoparathyroidism Type I); or in which these biochemical and skeletal features coexist but the cAMP response to exogenous PTH is normal (pseudohypoparathyroidism Type II) (Drezner, Neelon and Lebovitz, 1973). In neither of these conditions are the abnormalities in plasma phosphate or calcium corrected by exogenous PTH. Some of these conditions are familial. In some patients with pseudohypoparathyroidism, the response to exogenous PTH may be restored by vitamin D, with or without restoration of the cAMP response (Stogman and Fischer, 1975; Parfitt, 1976). Although patients with pseudohypoparathyroidism have the biochemical features of parathyroid insufficiency, the parathyroids themselves are overactive presumably in an attempt to overcome the hypocalcaemia.

PARATHYROID OVERACTIVITY

Primary hyperparathyroidism

Pathology

We need to consider the causes of parathyroid overactivity and their effect on target organs, particularly bone.

(a) The parathyroids
The pathology of the parathyroids is well dealt with by Williams (1974). Hyperparathyroidism itself may be due to adenoma or carcinoma, to hyperplasia or due to ectopic production of PTH.

The commonest cause of primary hyperparathyroidism is a parathyroid adenoma, which occurs in approximately 80 per cent of patients. The remaining 20 per cent comprise primary chief cell and water-clear cell hyperplasia, and carcinoma. The dominant problem in the pathology of the parathyroid is the separation of hyperplasia from neoplasia. Carcinoma may be difficult to diagnose histologically. Clinically, it gives a different picture from adenoma, sometimes with palpably enlarged parathyroids and often with recurrence. It has been suggested that parathyroid adenomata occasionally follow previous radiotherapy to the thyroid. Primary hyperplasia may occur in association with the MEA syndromes. Parathyroid hyperplasia is most commonly secondary to prolonged hypocalcaemia (p. 172).

(b) The bone
The characteristics of parathyroid bone disease are excessive osteoclastic bone resorption and fibrosis of the marrow. There is also evidence of increased rate of bone formation. Where bone turnover is very rapid considerable woven collagen is formed. Osteoid may also be increased in amount but this does not necessarily indicate a primary disorder of mineralization.

Although clinical bone disease is not common, quantitative histological studies show that the bone is nearly always abnormal (Byers and Smith, 1971). Bone disease tends to occur in patients with the biggest parathyroid tumours, and those with the most marked hypercalcaemia, and is presumably directly related to the amount of circulating parathyroid hormone. Clinical studies have shown that the bone disease and renal stones tend to occur in different patients, but there is no biochemical evidence for the suggestion that these differences reflect different types of parathyroid hormone. In some patients, which include those with parathyroid carcinoma, bone disease and renal stones may occur together.

Clinical features

Primary hyperparathyroidism is most common in women, particularly after the menopause and may present with a wide variety of symptoms. Paterson (1974) gives a good summary of these. Watson (1974) in a review of 100 consecutive patients (out of 350) found that the main features occurred in the following order: renal stones and nephrocalcinosis (47 per cent), osteitis fibrosa (13 per cent), gastrointestinal manifestations (12 per cent), hypercalcaemic symptoms (7 per cent), hypertension (40 per cent), association with other endocrine disorders (4 per cent) and psychiatric disorder (2 per cent). Analysis of other series (Tomlinson and O'Riordan, 1978) gives very similar results, with renal stones being far more frequent than bone disease. It is impressive how often psychological abnormalities, particularly depression, occur, although they may not be the presenting symptoms. Most recent series will also include patients who have been detected by biochemical screening.

Some of these presenting symptoms require amplification. Since renal calculi are the commonest and often the only symptoms of hyperparathyroidism, any patient with calculi should have the plasma and urine calcium measured. It has been estimated that between 3 to 10 per cent of patients with recurring renal calculi have hyperparathyroidism. Nephrocalcinosis itself may first be diagnosed radiologically rather than by its symptoms. Gastrointestinal symptoms include those of peptic ulceration (which is significantly associated with primary hyperparathyroidism) and of pancreatitis. Subacute and chronic pancreatitis may cause recurrent abdominal pain in hyperparathyroidism; and acute pancreatitis may lower the raised serum calcium to normal levels. Osteitis fibrosa may present in a number of ways; but classically there is bone pain and tenderness, deformity and pathological fractures. Often osteitis fibrosa is first discovered radiologically; clinically resorption of the ends of the phalanges may be noted with 'pseudo-clubbing'. Musculoskeletal symptoms may vary considerably (Patten et al., 1974); thus there may be difficulty in walking, stiffness of the joints, or proximal muscle weakness. The joint symptoms may be due to pyrophosphate arthropathy, to gout, or to a mechanical arthritis associated with the bone disease.

The symptoms of hypercalcaemia are nocturia, thirst, anorexia, vomiting, constipation and depression. The nocturia is due to loss of the ability to concentrate the urine because of the effect of hypercalcaemia on the renal tubule and is its first and most prominent effect. Depression is also generally considered to be due to the hypercalcaemia, but is a reasonable response to the other symptoms.

Physical examination of the patient with primary hyperparathyroidism may show very little. However corneal calcification (when present) (Chapter 2) is a most useful clue. The patient with hypercalcaemia often appears ill and may be dehydrated. Hypertension is commoner in hyperparathyroidism than in normal subjects. The bones may be tender on percussion. Very rarely, the tumour itself is palpable, and this is particularly so in the rare parathyroid carcinoma. In 61 patients with parathyroid carcinoma the presenting signs and symptoms were bone disease (39 patients), a palpable mass in the neck (19), and renal stones (18). Sexes were equally affected (Schantz and Castleman, 1973).

Biochemistry

The typical biochemical changes in primary hyperparathyroidism are hypercalcaemia and hypophosphataemia. Despite much writing on 'normocalcaemic' primary hyperparathyroidism, this is a rare problem which is usually encountered in the search for the cause of recurrent renal stones. In practice the main problem is the differential diagnosis of an established hypercalcaemia (Table 6.I). The

PARATHYROID OVERACTIVITY

TABLE 6.I

Some causes of hypercalcaemia

Frequency	Cause		Features
Common	Neoplasm (usually hydrocortisone sensitive)	Lung, breast etc.	Of underlying disease. Sometimes biochemistry of hyperpara-thyroidism with low plasma P
		Myeloma	High ESR. Abnormal plasma proteins. Bence Jones (light chain) proteinuria. Multiple osteolytic lesions (particularly skull)
	Hyperparathyroidism	including parathyroid carcinoma	Hydrocortisone insensitive
Rare	Vitamin D overdose		Sometimes symptoms of acute hypercalcaemia – sore eyes, thirst, vomiting
	Vitamin D sensitivity – sarcoidosis		Nephrocalcinosis, splenomegaly, hilar glands +. Precipitated by physiological doses of vitamin D or sunlight
	Immobilization		Especially in young people, or immobilized Paget's disease
	Thyrotoxicosis	Spontaneous or self-medication	Plasma P may be increased. Total hydroxyproline increased.
	Milk-alkali syndrome		Very rare. Alkali and milk for peptic ulcer
	Addison's disease (hypoadrenalism)		
	Thiazide diuretics		
	Lymphoma		
	Leukemia		
	Acute renal failure		

These are arranged in approximate order of frequency. The commonest cause of hypercalcaemia is neoplastic disease

biochemical changes can be understood in terms of the physiology of the hormone. Thus the hypercalcaemia results from autonomous overactivity of the parathyroid tumour (usually an adenoma), with increased intestinal absorption of calcium, increased bone resorption and increased renal-tubular reabsorption of calcium. Since the excretion of calcium is related to its filtered load, hyper-calcaemia causes hypercalcuria. In primary hyperparathyroidism this increase in urine calcium is less than that produced by artificial elevation of plasma calcium in normal subjects and in patients with hypercalcaemia due to other causes because PTH also increases tubular resorption of calcium. The hypophosphataemia is due to a reduction in renal-tubular reabsorption, that is the TmP/GFR.

The plasma alkaline phosphatase is normal where there is no radiological or clinical evidence of bone disease. Therefore it is normal in the majority of patients with primary hyperparathyroidism. In those with bone disease it correlates with the urinary total hydroxyproline and roughly with the radio-

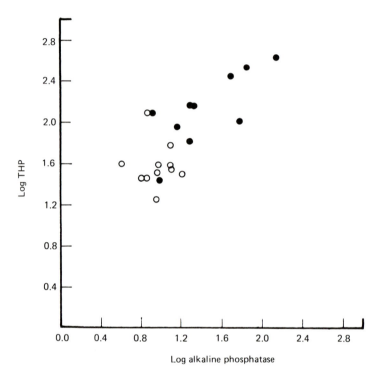

Figure 6.3. The correlation between plasma alkaline phosphatase, total hydroxyproline and clinical bone disease in patients with primary hyper-parathyroidism. Bone disease has been assessed radiologically, • with bone disease, ○ without. (From Byers and Smith, 1971)

logical extent of the bone disease (*Figure 6.3*). Other biochemical abnormalities which may sometimes be of use in the diagnosis are the presence of a hyper-chloraemic acidosis due to a defect in renal-tubular reabsorption of bicarbonate. Although consideration of multiple biochemical values may be useful in separating hypercalcaemic patients with hyperparathyroidism from those without (*see* Reeve, 1977), Tomlinson and O'Riordan (1978) are of the opinion that changes in phosphate clearance or in plasma chloride and bicarbonate may be of little diagnostic value. Patients with clinical bone disease also tend to have a reversible generalized amino-aciduria. Finally hyperuricaemia may be a clinical clue.

Radiology

This may help with the diagnosis of the bone disease, and with localization of the tumour (*see also* p. 170). Subperiosteal resorption of the phalanges is diagnostic of parathyroid overactivity (primary, secondary or tertiary); it is said to occur most markedly on the radial border of the proximal and middle phalanges of the

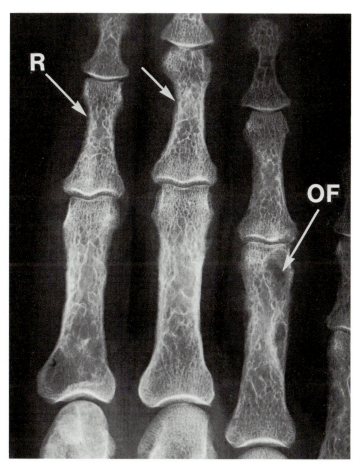

Figure 6.4. The appearance of the phalanges in a patient with primary hyperparathyroidism, demonstrating subperiosteal resorption (R) and localized areas of osteitis fibrosa cystica (OF)

index and middle finger. It is present only in a minority of patients with primary hyperparathyroidism (in general those with a raised alkaline phosphatase). Resorption and consequent shortening, may also affect the distal phalanges but this may occur in other diseases such as the acro-osteolysis of polyvinyl chloride workers (Harris and Adams, 1967). Resorption may be found in other places such as the medial borders of the upper tibiae, symphysis pubis, femoral necks, and outer ends of the clavicles. Occasionally more localized lesions are seen, with 'cysts' in the ribs or other bones (*Figure 6.4*), and the skull may be affected with a 'pepper-pot' appearance or with larger translucent areas. The lamina dura of the teeth may be absent in some patients but this is of little diagnostic value since it may occur in several other conditions (*see* Paterson, 1974). Other non-diagnostic changes include skeletal rarefaction, suggesting osteoporosis. A clue may be provided by articular chondrocalcinosis, which is said to be more frequent in hyperparathyroidism than in the normal population. Very occasionally the adenoma itself may be recognized by producing a deformity on a barium swallow.

Bone biopsy

This is not routinely necessary in diagnosis but has been widely used in research. A problem may arise where a coexistent bone disease, such as Paget's disease, is suspected: or where, for instance, the x-rays suggest secondary neoplastic deposits or myeloma.

Immunoassay

It is not possible to measure the normal chemical amount or biological activity of PTH directly in blood, but radio-immunoassay has proved to be very useful. However different workers have used different standards, and several fragments of PTH exist in plasma which are detected by some sera and not others. There is an overlap between the increased values of immunoreactive parathyroid hormone (iPTH) found in primary hyperparathyroidism, and the values in normal subjects, and the values for iPTH must be interpreted in relation to the plasma calcium. This is because hypercalcaemia suppresses the secretion of parathyroid hormone. Thus in a hypercalcaemic patient with primary hyperparathyroidism the concentration of iPTH should be increased, but in patients with hypercalcaemia due to other causes it should be less than normal or undetectable. In pseudohyperparathyroidism (p. 170) associated with cancer, iPTH levels are also often normal or reduced. In renal failure iPTH levels are increased. Examples are given by Peacock (1976) (*Figure 6.5*), and by Tomlinson and O'Riordan (1978). The last authors quote a normal upper limit of 1.0 ng/ml and a lower limit of detection of 0.15 ng/ml.

Diagnosis and differential diagnosis

The steps in the diagnosis of primary hyperparathyroidism are to establish that sustained hypercalcaemia exists, to distinguish it from other causes of hypercalcaemia, to demonstrate, if necessary by immunoassay, that there is over-activity of the parathyroids, and (again if necessary) to localize the tumour preoperatively. Immunoassay is being increasingly used in the diagnosis of primary hyperparathyroidism and this is logical. However it is usually possible to make the diagnosis without it and large numbers of patients have been success-fully treated before such assays were available. Similarly the localization of a tumour is not necessary as a routine before neck exploration, but can be very useful where such an exploration has failed. Raisz (1971) wrote a useful account of the diagnosis of hyperparathyroidism without an immunoassay.

There are many recognized causes of hypercalcaemia, of which neoplasm is the most important (Table 6.I), and should always be considered and excluded clinically. In those patients with primary hyperparathyroidism, with hypercalcaemia, hypophosphataemia, hyperphosphatasia and radiological evidence of osteitis fibrosa, and without clinical evidence of neoplasm, no further investigation is needed. However only a minority of patients with hyper-parathyroidism have clinical bone disease and in the majority further differen-tiation from other causes of hypercalcaemia (Table 6.I) is necessary. Unless there

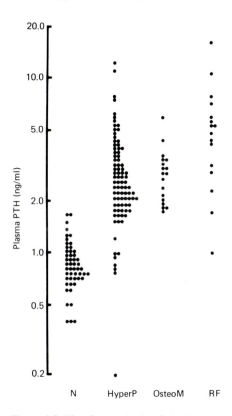

Figure 6.5. The changes in circulating immuno-reactive parathyroid hormone (iPTH) *in different conditions.* R.F. = *Renal failure.*
(From Peacock, 1976)

is obvious evidence of sarcoidosis, thyrotoxicosis, neoplasm or vitamin D over-dosage, and the patient is not on thiazide diuretics it is still useful to do a hydrocortisone suppression test (Dent and Watson, 1968). The way in which this test works is unknown. In its original form hydrocortisone was given at a dose of 40 mg 8-hourly for 10 days and blood taken at days 0, 5, 8 and 10. If the plasma calcium remains elevated, this is strongly in favour of primary hyper-parathyroidism. It is important to correct the plasma calcium for the change in plasma protein or albumin concentration (p. 49) since hydrocortisone causes fluid retention and therefore may alter the plasma protein concentration.

However the plasma calcium may not fall (partly or completely) where there is a clinically obvious neoplasm. This test has been modified to one which uses prednisone, which is probably just as useful — however there are the theoretical objections to the use of synthetic corticosteroid derivatives, especially where there may be hepatic involvement as in hypercalcaemic sarcoidosis.

Hypercalcaemia in association with neoplastic disease may occur with or without bone metastases (Buckle, 1974). In the majority of patients with malignant disease and hypercalcaemia there are secondary deposits in bone, with local destruction. The factors which cause this local destruction probably include prostaglandins (Chapter 1). Carcinoma of the breast is the commonest tumour responsible for hypercalcaemia. Osteolytic activity of tumour tissue can be demonstrated *in vitro* and may be suppressed by aspirin or indomethacin. Similar findings have been reported with myeloma, where hypercalcaemia is approximately related to the degree of bone involvement.

The occurrence of hypercalcaemia without obvious metastases in the bone presents interesting clinical and biochemical problems. It may be that the metastases are present but not clinically detectable, in which case the bone scan or urinary hydroxyproline may help, or that the tumour is producing an ectopic PTH-like polypeptide. The commonest tumour that does this is squamous-cell carcinoma of the lung. Since the biochemistry is similar to that of primary hyperparathyroidism, particularly with a low plasma phosphate, this condition has been called pseudohyperparathyroidism (which is confusing). Although some patients may appear to have an increase in PTH-like peptides (Buckle, 1974), in others multiple immunoassays for proparathyroid hormone and PTH fragments as well as PTH may fail to detect any increase in the peripheral blood or tumour tissue (Powell *et al.*, 1973). Many other factors, which again include prostaglandins (p. 31) and also osteoclastic-stimulating factors may be responsible. Differentiation of pseudohyperparathyroidism from primary hyperparathyroidism may be difficult because of the similarity of the biochemical findings and even of the bone histology. However, in practice, it is usual to find low levels of iPTH in malignant hypercalcaemia. Of the rarer causes of hypercalcaemia (Table 6.I) self-medication with vitamin D is recorded.

Preoperative localization

Although it is not necessary to attempt routine localization of a parathyroid tumour, this may be important in the case of a re-exploration. Such methods as selenomethionine-scanning and inferior-thyroid arteriography have not proved reliable, but the measurement of iPTH levels in blood obtained by selective catheterization of veins draining the tumour may be very useful (Davies *et al.*, 1973; Bilezikian *et al.*, 1973).

Treatment

Orthodox teaching is that the only treatment of primary hyperparathyroidism is surgical, but this should not be accepted without thought, and it is doubtful whether one should operate on asymptomatic patients with mild hypercalcaemia (less than 3 mmol/l). The operation should involve removal of the adenoma

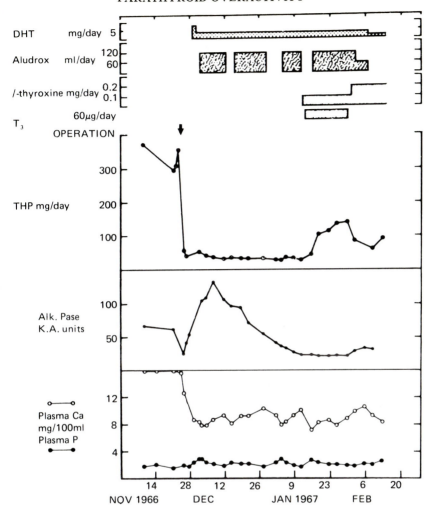

Figure 6.6. Biochemical changes after total parathyroidectomy and thyroidectomy in a patient with parathyroid carcinoma. The precipitous fall in urine total hydroxyproline (THP) *represents sudden cessation of bone resorption.* THP *excretion increases when thyroxine replacement is given. For further details see Smith (1969)*

together with identification and examination of the remaining three glands. Injection of toluidine blue or methylene blue at the time of operation may help to identify the adenoma. It is necessary to be aware of the clinical and biochemical changes which may occur following the operation, which will be most marked and prolonged in the patient with severe osteitis fibrosa.

Immediately after parathyroidectomy bone resorption falls together with a fall in iPTH levels. Plasma and urine hydroxyproline fall to normal levels (*Figure 6.6*). Bone formation, indicated by alkaline phosphatase (and possibly the non-dialysable fraction of urine hydroxyproline) continues. Hypocalcaemia is usually most marked by the third postoperative day, and in patients without osteitis fibrosa the plasma calcium may be normal again within a few days. However where bone disease is extensive, hypocalcaemia may be profound and persistent. This has been attributed to rapid bone healing ('hungry bones') and requires vigorous treatment, since the hypocalcaemia can lead to fits which in turn can cause the softened bones to fracture. Postoperative hypocalcaemia can be partly prevented by giving high doses (up to 4 mg daily) of vitamin D, either as D_2 or DHT, preoperatively. Postoperatively vitamin D should be continued together with large amounts of oral calcium, intermittent intravenous calcium to control symptoms of hypocalcaemia, and aluminium hydroxide, which lowers plasma phosphate and tends to increase plasma calcium. Rarely hypomagnesaemia may occur (Davies and Friedman, 1966).

It was previously necessary to give vitamin D preoperatively since it took some days to have its full effect, but by modifying the hypocalcaemic effect of parathyroidectomy this unfortunately abolished the biochemical demonstration that the operation had been successful. The postoperative hypocalcaemia can now be managed much more easily by using a vitamin D metabolite such as $1\alpha(OH)D$, which need not be given until after the neck exploration, acts quickly and does not accumulate.

Although recovery from parathyroidectomy is usually uneventful, pancreatitis or gout can occur. The aims of parathyroidectomy are to correct hypercalcaemia and to prevent progressive renal failure, which seems to be the main cause of death. There is evidence that parathyroidectomy succeeds in this and may also improve the associated hypertension (Rosenthal and Roy, 1972). In patients with mild hyperparathyroidism surgery is not necessarily indicated. In post-menopausal women the plasma and urine calcium may be reduced by oestrogens. After unsuccessful neck exploration persistent hypercalcaemia has been treated (for short periods at least) with oral or intermittent intravenous phosphate, and hypercalcaemia due to parathyroid carcinoma with mithramycin or calcitonin. The emergency treatment of hypercalcaemia (Paterson, 1974) is outside the scope of this book.

Multiple endocrine adenoma syndromes

Overactivity of the parathyroids may be familial and it may also be associated with overactivity of other endocrine organs. The situation is complex (Marx *et al.*, 1977) but two particular associations are recognized. Type I multiple endocrine adenoma (MEA) syndrome includes hyperparathyroidism, pituitary adenomata, insulin and gastrin-secreting tumours of the pancreas and gastric hyperacidity (Zollinger–Ellison syndrome): Type II MEA syndrome, also known as Sipples' syndrome (Keiser *et al.*, 1973), includes hyperparathyroidism, medullary carcinoma of the thyroid and phaeochromocytoma. It has been proposed that these associations occur because the cells have a common origin from the neural crest, and form part of a general endocrine system. The MEA syndromes may also be related to the endocrine systems of the gut, the so-called APUD system (Pearse, 1973). Familial neonatal hyperparathyroidism is a separate and rare condition occurring in siblings.

Secondary (and tertiary) hyperparathyroidism

Where hypocalcaemia is prolonged, most often due to renal glomerular failure or to gluten-sensitive enteropathy the parathyroid glands respond by increasing both their activity and size in an attempt to restore the plasma calcium towards normal. Thus to the clinical features of the underlying disorder are added those of secondary hyperparathyroidism with excessive resorption of bone. Particularly in renal disease this may cause rapid bone loss and deformity especially in growing subjects, where the combination of a disorder of mineralization and resorption of bone may produce a 'rotting stump' appearance on x-ray (Chapter 4). It is characteristic of secondary hyperparathyroidism that the parathyroid overactivity should decline when the underlying disease is treated, and this is best seen in the treatment of osteomalacia due to vitamin D deficiency and malabsorption (Tomlinson and O'Riordan, 1978, Figure 9).

There are however situations where hypercalcaemia develops either without treatment, or when vitamin D is given, and persists. This was initially described in patients with steatorrhoea and osteomalacia, and it was subsequently proposed, with considerable evidence, that in such patients one or more of the hyperplastic parathyroid glands had become autonomous with the production of 'tertiary' hyperparathyroidism (Davies, Dent and Watson, 1968). Subsequently the causal relation between parathyroid adenomata and prolonged hypocalcaemia was queried, and it was noted that patients with primary hyperparathyroidism might be vitamin-D-deficient. The situation in the hypercalcaemia after renal transplantation is complex, and suppression of the hyperplastic glands may take a long time because of their size.

HYPOPARATHYROIDISM

Parathyroid insufficiency may occur after surgical removal of the parathyroids, in idiopathic hypoparathyroidism, and in a familial form of hypoparathyroidism which is often associated with manifestations of auto-immune disease, including moniliasis, malabsorption, thyroid and adrenal failure and pernicious anaemia. In such patients the levels of iPTH are undetectably low but the cAMP response to exogenous PTH is maintained (*Figure 6.7*). This distinguishes parathyroid insufficiency from pseudohypoparathyroidism (PHP), in which the biochemical features of hypoparathyroidism are associated with characteristic skeletal abnormalities. PHP is inherited and in the most common form the production of cAMP is defective. Hypocalcaemia leads to compensatory overactivity and enlargement of the parathyroids. Thus the cAMP response to exogenous PTH is absent or blunted (*Figure 6.7*) whilst the circulating level of iPTH is high. In PHP there are no demonstrable changes in calcitonin levels and so far as is known the PTH is normal in structure and biologically active. Variations of PHP appear to exist, and disorders are described in which the cAMP is present but there is still end-organ resistance (PHP Type II), and also where the cAMP response is restored by giving vitamin D. Patients who have the skeletal manifestations of PHP but with normal biochemistry may be found in families with PHP, and to them the term pseudopseudohypoparathyroidism (PPHP) is applied.

Accidental removal of, or damage to, the parathyroids following thyroid surgery is probably still the commonest cause of parathyroid insufficiency. The time of onset of symptoms will vary with the cause, and idiopathic hypoparathyroidism may present in infancy or childhood (a rare cause of neonatal hypoparathyroidism is maternal hyperparathyroidism). Of the numerous clinical

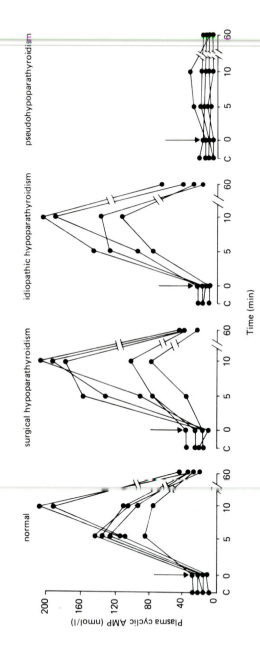

Figure 6.7. Demonstrates the difference in cAMP response to exogenous PTH in different types of hypoparathyroidism. This records the changes in plasma cAMP; similar but less rapid changes are seen in urine cAMP. (From Tomlinson and O'Riordan, 1978)

features, only some can be explained (*Figure 6.8*). Hypoparathyroidism should always be excluded in a child with unexplained epilepsy. Prolonged hypocalcaemia may lead to dementia as well as depression. The ectopic calcification which occurs in the basal ganglia and subcutaneous tissues (*Figure 6.9*) may be partly due to

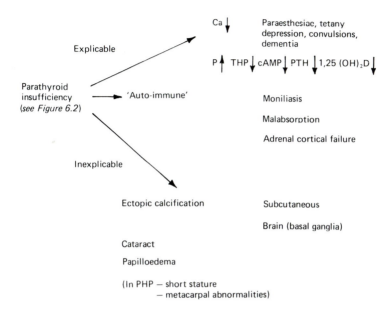

Figure 6.8. Explicable and inexplicable features of parathyroid insufficiency

the hyperphosphataemia, but since it is seen in patients with PPHP whose biochemistry is normal this cannot be the only cause; and the cause of cataracts and papilloedema is unknown. Skeletal abnormalities and mental simplicity are mainly features of PHP.

Diagnosis

In a patient with hypocalcaemia and hyperphosphataemia it is necessary to exclude renal-glomerular failure, occasionally steatorrhea, and PHP. The reason why some patients with malabsorption have biochemical features of hypoparathyroidism and appear to be resistant to the effect of PTH is not fully understood, although responsiveness to PTH may sometimes be restored by giving magnesium. The important distinction from PHP is based on the levels of iPTH and the cAMP response to exogenous PTH. In so-called pseudo-idiopathic hypoparathyroidism (p. 162) the situation is not clear.

Treatment

The accepted treatment of hypoparathyroidism with large doses of vitamin D is not without hazard, since it may be difficult to control its dose and intermittent

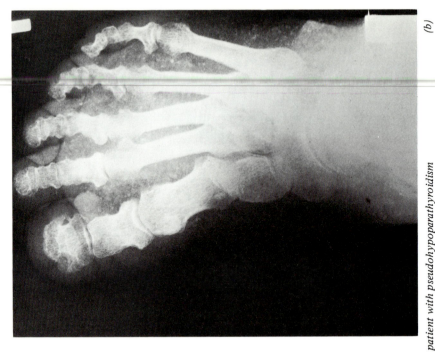

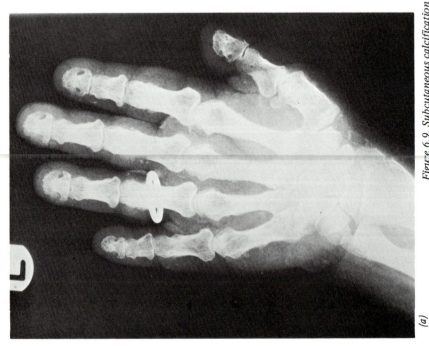

Figure 6.9. Subcutaneous calcification in a patient with pseudohypoparathyroidism

vitamin D toxicity occurs. The usual daily dose of calciferol or DHT is around 0.5 mg (20 000 iu.). The lack of parathyroid hormone leads to reduced activity of the renal-1-hydroxylase, which could account for the apparent vitamin D resistance. Recorded plasma levels of $1,25(OH)_2D$ are low. It has now been shown that hypoparathyroidism may be rapidly and satisfactorily treated with $1,25(OH)_2D$ or $1\alpha(OH)D$ in physiological doses (Russell *et al.*, 1974, *Figure 6.10;* Davies *et al.*, 1977).

PSEUDOHYPOPARATHYROIDISM (PHP)

The clinical picture is similar to that of hypoparathyroidism, but with additional features, which includes family history, skeletal abnormalities and mental simplicity. The exact form of inheritance is not known and the expression is variable, but existing evidence suggests inheritance as an X-linked dominant

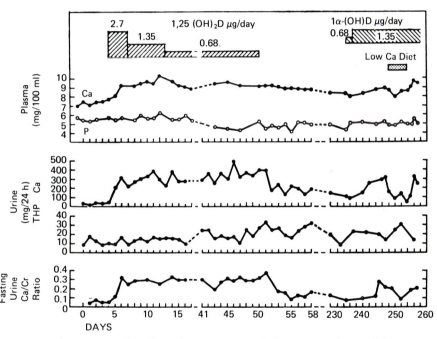

Figure 6.10. To show the effect of vitamin D metabolites in a patient with hypoparathyroidism. There is a rapid increase in both the plasma and urine calcium and also in the fasting calcium/creatinine ratio in the urine. The changes are more likely to be due to an increase in calcium absorption than to bone resorption, and there is little change in the urine THP excretion. (For further details see Russell et al., *1974)*

(Potts, 1978). The general appearance is usually abnormal with short stature, short neck and round face. Although many skeletal abnormalities are described, the most typical is short fourth or fifth metacarpals (or metatarsals). The bones may be unusually dense (*Figure 6.9*). Occasionally patients with the biochemistry of hypoparathyroidism show evidence of excessive bone resorption (hypohyperparathyroidism). Mental backwardness is common and is likely to be a definite part of the syndrome rather than due to the effects of prolonged hypocalcaemia. Potts (1978) points out that mild hypothyroidism can occur, and that defects in taste and smell exist.

The treatment of PHP is similar to that of hypoparathyroidism, and the 1α-hydroxylated metabolites of vitamin D are effective.

SUMMARY

Primary hyperparathyroidism with bone disease is comparatively rare, but study of the biochemistry and effects of parathyroid hormone has considerably increased our knowledge of the mechanism of hormone action and of calcium metabolism. In contrast hypercalcaemia is a common problem and diagnosis of its cause is important. Recent work has shown that agents other than parathyroid hormone may stimulate bone resorption, and hypercalcaemia with neoplastic disease may be due to the effects of local factors such as prostaglandins. Further study of neoplastic hypercalcaemia is likely to lead to physiological and clinical advance. Study of the rare forms of parathyroid insufficiency has suggested blocks at various stages in the synthesis and action of PTH on its target organs.

REFERENCES

Arnaud, C. D. (1973). Parathyroid hormone: coming of age in clinical medicine. *Am. J. Med.* **55**, 577–581

Aurbach, G. D. and Potts, J. T. (1967). Parathyroid hormone. *Am. J. Med.* **42**, 1–8

Bilezikian, J. P., Doppman, J. L., Shimkin, P. M., Powell, D., Wells, S. A., Heath, D. A., Ketcham, A. S., Monchik, J., Mallette L. E., Potts, J. T. and Aurbach, G. D. (1973). Preoperative localization of abnormal parathyroid tissue. *Am. J. Med.* **55**, 505–514

Buckle, R. (1974). Ectopic PTH syndrome, pseudohyperparathyroidism, hypercalcaemia of malignancy. *Clinics Endocr. Metabolism* **3**, 237–251

Butcher, R. W. (1968). The role of cyclic AMP in hormone actions. *New Engl. J. Med.* **279**, 1378–1384

Byers, P. D. and Smith, R. (1971). Quantitative histology of bone in hyperparathyroidism *Q. Jl Med.* **40**, 471–486

Clinics in Endocrinology and Metabolism (1974). Disorders of the parathyroid glands. **3**, No. 2

Davies, D. R. and Friedman, M. (1966). Complications after parathyroidectomy. Fractures from low calcium and magnesium convulsions. *J. Bone Jt Surg.* **48B**, 117–126

Davies, D. R., Dent, C. E. and Watson, L. (1968). Tertiary hyperparathyroidism. *Br. med. J.* **3**, 395–399

Davies, D. R., Shaw, D. G., Ives, D. R., Thomas, B. M. and Watson, L. (1973). Selective venous catheterization and radio-immunoassay of parathyroid hormone in the diagnosis and localization of parathyroid tumours. *Lancet* **1**, 1079–1082

Davies, M., Hill, L. F., Taylor, C. M. and Stanbury, S. W. (1977). 1,25-hydroxycholecalciferol in hypoparathyroidism. *Lancet* **1**, 55–58

Dent, C. E. and Watson, L. (1968). The hydrocortisone test in primary and tertiary hyperparathyroidism. *Lancet* **2**, 662–664

Drezner, M., Neelon, F. A. and Lebovitz, H. E. (1973). Pseudohypoparathyroidism Type II: a possible defect in the reception of the cyclic AMP signal. *New Engl. J. Med.* **289**, 1056–1060

Fourman, P. and Royer, P. (1968). *Calcium Metabolism and the Bone.* 2nd edition. Oxford: Blackwells Scientific Publications

Goldman, L., Gordan, G. S. and Roof, B. S. (1971). The parathyroids; progress, problems and practice. *Curr. Probl. Surg.* August 1971. Chicago: Yearbook Medical Publishing Inc.

REFERENCES

Harris, D. K. and Adams, W. G. F. (1967). Acro-osteolysis occurring in men engaged in the polymerisation of vinyl chloride. *Br. Med. J.* **3**, 712–714

Hunter, D. and Turnbull, H. M. (1931). Hyperparathyroidism: Generalized osteitis fibrosa. With observations upon the bones, the parathyroid tumours and normal parathyroid glands. *Br. J. Surg.* **19**, 203–284

Keiser, H. R., Beaven, M. A., Doppman, J., Wells, S. and Buja, L. M. (1973). Sipples' syndrome. Medullary thyroid carcinoma, phaeochromocytoma and parathyroid disease. *Ann. inter. Med.* **78**, 561–579

Marx, S. J., Spiegel, A. M., Brown, E. M. and Aurbach, G. D. (1977). Family studies in patients with primary parathyroid hyperplasia. *Am. J. Med.* **62**, 698–706

Nusynowitz, M. L. and Klein, M. H. (1973). Pseudo-idiopathic hypoparathyroidism. Hypoparathyroidism with ineffective parathyroid hormone. *Am. J. Med.* **55**, 677–686

O'Riordan, J. L. H. (1972). Parathyroid hormone and hyperparathyroidism. In *Advanced Medicine*. Vol. 8. Ed. G. Neale. pp. 22–39. London: Pitman Medical

Parfitt, A. M. (1976). The actions of parathyroid hormone on bone; relation to bone remodelling and turnover, calcium homeostasis and metabolic bone disease. Parts I–IV. *Metabolism* **25**, 809–844, 909–955, 1033–1069, 1157–1188

Paterson, C. R. (1974). *Metabolic Disorders of Bone*. Oxford: Blackwells Scientific Publications

Patten, B. M., Bilezikian, J. P., Mallette, L. E., Prince, A., Engel, W. K. and Aurbach, G. D. (1974). Neuromuscular disease in hyperparathyroidism. *Ann. intern. Med.* **80**, 182–193

Peacock, M. (1976). Parathyroid hormone and calcitonin in calcium, phosphate and magnesium metabolism. Ed. B. E. C. Nordin. London and Edinburgh: Churchill Livingstone

Pearse, A. G. E. (1973). The gut as an endocrine organ. In *Advanced Medicine*. Vol. 9. Ed. G. Walker. pp. 400–409. London: Pitman Medical

Potts, J. T. (1978). Pseudohypoparathyroidism. In *Metabolic Bases of Inherited Disease*. Eds J. B. Stanbury, J. B. Wyngaarden and A. S. Fredrickson. pp. 1350–1365. New York: McGraw Hill

Potts, J. T. and Deftos, L. J. (1974). Parathyroid hormone, thyrocalcitonin, Vitamin D bone and bone mineral metabolism. In *Duncan's Diseases of Metabolism*. Eds P. K. Bondy and L. E. Rosenberg. 7th edn. pp. 1225–1430. London: W. B. Saunders Co.

Powell, D., Singer, F. R., Murray, T. M., Minkin, C. and Potts, J. T. (1973). Nonparathyroid humoral hypercalcaemia in patients with neoplastic diseases. *New Engl. J. Med.* **289**, 176–181

Raisz, L. G. (1971). The diagnosis of hyperparathyroidism (or what to do until the immuno-assay comes). *New Engl. J. Med.* **285**, 1006–1009

Reeve, J. (1977). Disorders of plasma calcium. *Hospital Update* **3**, 19–30

Robinson, C. J. (1974). The physiology of parathyroid hormone. *Clinics Endocr. Metabolism* **3**, 389–421

Rosenthal, F. D. and Roy, S. (1972). Hypertension and hyperparathyroidism. *Br. med. J.* **4**, 396–397

Russell, R. G. G., Smith, R., Walton, R. J., Preston, C., Basson, R., Henderson, R. G. and Norman, A. W. (1974). 1,25-dihydroxycholecalciferol and 1α-hydroxycholecalciferol in hypoparathyroidism. *Lancet* **1**, 14–17

Schantz, A. and Castleman, B. (1973). Parathyroid carcinoma. A study of 70 cases. *Cancer* **31**, 600–605

Smith, R. (1969). Dissociation between changes in urinary total hydroxyproline and plasma alkaline phosphatase after removal of parathyroid tumour. *Clinica chim. Acta* **23**, 421–426

Stogman, W. and Fischer, J. (1975). Pseudohypoparathyroidism. Disappearance of resistance to parathyroid extract during treatment with vitamin D. *Am. J. Med.* **59**, 140–144

Tomlinson, S. and O'Riordan, J. L. H. (1978). The parathyroids. *Br. J. Hosp. Med.* **19**, 40–53

Vaughan, J. (1975). *The Physiology of Bone*. 2nd edition. Oxford: Clarendon Press

Watson, L. (1974). Primary hyperparathyroidism. *Clinics Endocr. Metabolism* **3**, 215–235

Williams, E. D. (1974). Pathology of the parathyroid glands. *Clinics Endocr. Metabolism* **3**, 285–301

7

Osteogenesis Imperfecta

INTRODUCTION

Osteogenesis imperfecta, probably the commonest of the inherited disorders of connective tissue which primarily affect the skeleton, is about as rare as haemophilia and one-third as common as Duchenne muscular dystrophy (Wynne-Davies, 1973). Nevertheless, it deserves our attention because of the severe crippling it produces and because its cause remains unknown.

The features of this ancient disease (Gray, 1969) which were described nearly two centuries ago (Ekman, 1788) suggest an abnormality of collagen. They do however vary from one individual to another which makes logical classification difficult. Indeed it is likely that the brittle bones which give osteogenesis imperfecta its name are the manifestation of a number of separate disorders. Current advances in biochemistry of connective tissue (Chapter 1) should help to settle this point.

Patients with osteogenesis imperfecta are seen by a variety of specialists, particularly orthopaedic surgeons, physicians and paediatricians, and their care tends to be fragmented. Recent reviews of significant numbers of cases are therefore useful and provide some guide to prognosis and treatment (King and Bobechko, 1971; Falvo, Root and Bullough, 1974; Bauze, Smith and Francis, 1975).

Table 7.I gives the main points of some selected papers. Earlier references, back to 1678, are recorded in a useful paper by Shoenfeld, Fried and Ehrenfeld (1975).

The clinical aspects of osteogenesis imperfecta are important and will be dealt with first.

CLINICAL PICTURE

The main clinical features of osteogenesis imperfecta are fragile bones, dwarfism, blue sclerae, dentinogenesis imperfecta and deafness. Hypermobility may be seen and reviews refer also to poor skin healing, excessive bleeding, muscle weakness

TABLE 7.I
Some studies on osteogenesis imperfecta

Author	Origin	Observation
Clinical		
Caniggia, Stuart and Guideri (1958)	Italy	4 families. 23 affected cases. Consider all forms the same disease
Ibsen (1967)	California	Review of reported cases. Suggest distinct varieties including some with normal-coloured sclerae
King and Bobechko (1971)	Toronto	60 cases. Orthopaedic review, with results of correction of deformity
Falvo, Root and Bullough (1974)	New York	90 cases with 12 'osteogenesis imperfecta congenita'. Importance of bowing of long bones
Bauze, Smith and Francis (1975)	Oxford	42 patients. Bone deformities and white sclerae associated
Biochemistry		
Solomons and Styner (1969)		Serum and urine pyrophosphate increased
Russell *et al.* (1971)		Plasma pyrophosphate normal
Langness and Behnke (1971)		Increased total urine hydroxyproline and plasma hypoprotein
Smith, Francis and Bauze (1975)		Normal urine hydroxyproline for age
Pathology		
Doty and Matthews (1971)		Electron microscopy of bone in 'osteogenesis imperfecta tarda' suggests primary abnormality of osteoblasts
Falvo and Bullough (1973)		Increased woven bone particularly in 'osteogenesis imperfecta congenita', and increased numbers of osteocytes
Riley, Jowsey and Brown (1973)		Normal ultrastructure of osteocytes, bone collagen, scleral collagen. Diameter corneal collagen fibres reduced
Treatment		
Castells *et al.* (1972)		Porcine calcitonin
Williams *et al.* (1973)		Surgical treatment
Granda, Falvo and Bullough (1977)		Magnesium oxide
Rosenberg *et al.* (1977)		Long-term salmon calcitonin. Results variable

TABLE 7.I (contd)

Author	Origin	Observation
	Cause	
Francis, Smith and Bauze (1974)		Instability of polymeric dermal collagen in severe osteogenesis imperfecta
Dixon, Millar and Veis (1975)		Quantity and composition of non-collagenous organic matrix abnormal
Penttinen *et al.* (1975)		Fibroblasts from an infant with severe osteogenesis imperfecta synthesized more Type III collagen (compared to Type I) than normal
Sykes, Francis and Smith (1977)		Altered relation of Type III and Type I collagen in skin
Fujii, Kajiwari and Korosu (1977)		Altered content Type III collagen and proportion of cross-links in skin
Trelstad, Rubin and Gross (1977)		Increased hydroxylation of lysine, particularly in bone collagen

and excessive sweating. A division may conveniently be drawn according to the absence or presence of long-bone deformity between those with mild and those with severe bone disease.

Classification

The established classifications of patients with osteogenesis imperfecta recognize a severe sporadic form with intra-uterine fractures and early death, osteogenesis imperfecta 'congenita', and milder often dominantly inherited forms in whom fractures may be present at birth or within the first year of life ('tarda gravis'), or after the first year (tarda levis'). The detailed classifications need not concern us further, but it is often considered that babies born with intra-uterine fractures ('osteogenesis imperfecta congenita') will die at birth or soon after. This may be so, but a review (Bauze, Smith and Francis, 1975) of 42 Oxford patients (*Figure 7.1*) included a significant number of surviving children and adults who had had fractures at birth (and possibly before) and this is a feature of other reports. This work also agreed with a New York study (Falvo, Root and Bullough, 1974) that the degree of bone deformity was a useful basis for classifying the severity of osteogenesis imperfecta. Apart from the practical result that severely deformed patients are unlikely to be able to walk, the presence of deformity may also have pathological and biochemical significance.

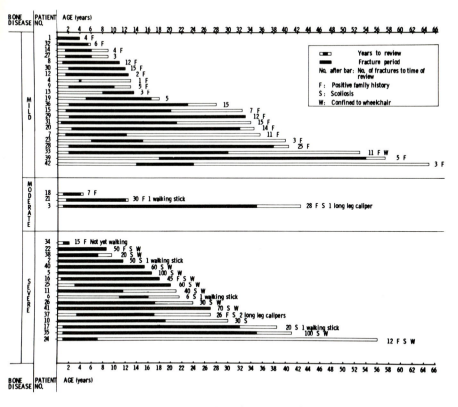

Figure 7.1. Demonstrates the individual clinical course of 42 patients with osteogenesis imperfecta. In this study patients were grouped according to the presence ('severe') or absence ('mild') of obvious long-bone deformity. Note that: (1), Nearly all patients who had fractures at birth were subsequently severely disabled, but patients who developed severe bone disease did not necessarily have fractures at birth; (2), Scoliosis and severe bone disease were closely associated; and (3), Fractures commonly continued after puberty.
(From Bauze, Smith and Francis, 1975)

Incidence and genetics

The incidence of this condition is said to be between 1 in 50 000 births for the 'congenita' type to 1 in 25 000 living for the 'tarda' form (Wynne-Davies, 1973).

Patients with mild osteogenesis imperfecta and blue sclerae transmit this disorder in a dominant fashion from one generation to the next. In the Oxford patients no parents with mild disease produced severely affected infants with fractures at birth. Infants with severe osteogenesis imperfecta with intra-uterine fractures occur sporadically, and the parents are apparently normal; it is rare for siblings to be affected, which could argue against recessive inheritance and in favour of a mutation. The genetic make up of offspring of an adult with 'osteogenesis imperfecta congenita' or two parents with mild osteogenesis imperfecta are not well known (*see, however,* Suen, Harris and Berman, 1974). The inheritance of some sub-groups of severe osteogenesis imperfecta such as those with hyperplastic callus, may differ from that of the severe group as a whole. Interesting pedigrees are discussed by McKusick (1972).

Mild osteogenesis imperfecta

In this type the onset of fractures may be delayed until childhood, and the number of fractures may be small. Fractures are said to decrease after childhood (*but see Figure 7.1*); they can occasionally become troublesome after the menopause, especially in the vertebrae (contributing to 'postmenopausal' osteoporosis).

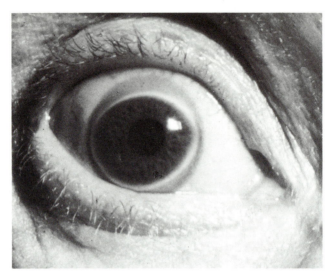

Figure 7.2. Arcus juvenilis (embryotoxon) in an adult with osteogenesis imperfecta. This appearance had been present since childhood

Although eventual stature is often reduced, fractures heal well with little or no deformity. The fact that the fractures are due to osteogenesis imperfecta is usually established by finding blue sclerae and/or brittle bones in previous generations, compatible with dominant inheritance. Where the sclerae are excessively blue this is usually attributed to their increased transparency (Eddowes, 1900), considered by Ruedemann (1953) to result from a reduction in collagen, which allows the pigmented coats of the choroid to be seen. This author recorded the pathological findings in two infants with fractures who died shortly after birth, and one adult who had fractures as a child. It can be difficult to assess the blueness of the sclerae (especially in the presence of eye-shadow). The type of blueness may differ and sclerae are normally slightly blue in babies. Moreover sclerae are not uniformly coloured and are whiter immediately near the limbus (*see* Nordin, 1973 — frontispiece Figure A). Other ocular features include arcus juvenilis (*Figure 7.2*) and occasionally keratoconus. Dentinogenesis imperfecta occurs in some patients (p. 186). Hearing is often affected, and the deafness resembles that due to otosclerosis, but there are many differences. A detailed account is given by Bergstrom (1977). It is interesting that the tympanic membranes may appear blue. In the cardiovascular system floppy or redundant

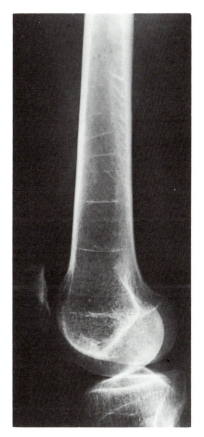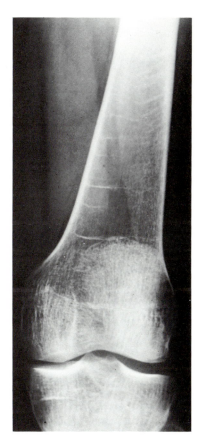

Figure 7.3. Mild abnormalities of bone structure in a patient from a large family with dominantly-inherited osteogenesis imperfecta

mitral valves, or aortic incompetence, are described. X-rays of the bones may show only very mild changes, with a thin cortex especially towards the ends of the long bones, with Harris lines and with oblique lines extending in from the cortex (*Figure 7.3*) and with some modelling defect. More severe changes without significant deformity can occur.

Severe osteogenesis imperfecta

The bones are excessively fragile and distorted and the repeated fractures often begin *in utero*. Gross deformity with shortening of the long bones is common. Disability and dwarfism are severe. These patients rarely walk, even with aids. There is usually no family history. In the Oxford series severe scoliosis appeared to be almost universal. At birth, examination shows shortness of the limbs (micromelia) (*Figure 7.4*) relative to the trunk. The skull is soft, with deficient bone formation (caput membranaceum); it may be enlarged or asymmetric with transverse widening; and, rarely, hydrocephalus can be a problem. Some changes in shape may be a mechanical result of the baby lying on its back and the soft head spreading out sideways. Characteristically the vault of the skull is large and broad with a temporal bulge and a small triangular face beneath it.

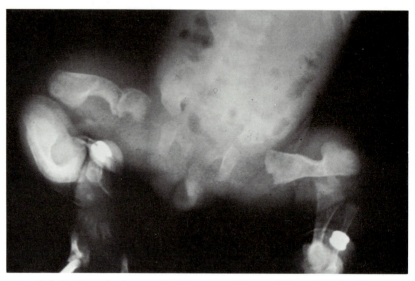

Figure 7.4. Radiograph of a neonate with severe osteogenesis imperfecta, to demonstrate micromelia. The bones of the leg are short, thick and bowed; there is an intra-uterine fracture of the right femur

In the majority of patients with severe osteogenesis imperfecta the teeth are abnormal due to a defect in the dentine (Rao and Witkop, 1971). Although the biochemistry and histology of such teeth is not well studied, this defect in dentine causes the teeth to appear opalescent with a brown or bluish discolouration. The enamel is not properly supported and tends to break off, and such teeth may wear down very quickly. Although it is likely that the dentine of all teeth is affected, clinically some teeth are often more affected than others. For instance, the permanent upper incisors may appear almost normal. Radiologically, affected teeth have reduction or obliteration of the pulp space, with reduced pulp canals and often small roots (*Figure 7.5*). It is interesting to note that some patients with severe bone disease may have clinically normal teeth. Levin, Salinas and Jorgenson (1978) suggest that on the basis of dental findings at least two autosomal dominant types of osteogenesis imperfecta exist.

At birth the sclerae are often noted to be blue but later may appear less blue or normal.

Patients with severe bone disease tend not to develop deafness. However they do show considerable hypermobility. Muscular weakness, affecting mainly the proximal muscles, is recorded.

The radiological appearances may be very abnormal (*Figures 7.6–7.11* and p. 191) according to the type of bone disease present.

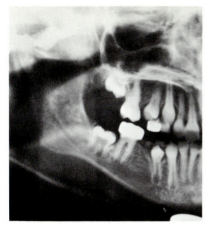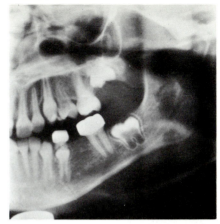

Figure 7.5. Panoramic radiograph of the teeth of a 9½-year-old boy with osteogenesis imperfecta. The roots are thin and the constriction at the junction of the crowns and roots is greater than normal. The pulp chambers and root canals of most teeth are narrow, although most of the developing mandibular second molars are wider than normal. (From Levin and Thompson, 1975)

The differences between those patients with mild and severe osteogenesis imperfecta (excluding a small intermediate group) in a series of 62 patients mainly from Oxford (Smith, Francis and Bauze, 1975) is shown in Table 7.II. It is compared with a similar survey in New York (Falvo, Root and Bullough, 1974) (Table 7.III). Patients with 'osteogenesis imperfecta congenita' in the New York series, who would come into our severe group were said to have blue sclerae, but at least some appear to have been examined very early in childhood.

Apart from long-bone deformity, the striking difference between mild and severe disease is in the incidence of scoliosis. The Toronto series (King and Bobechko, 1971) cannot be directly compared because of the different basis for classification. Again it is notable that 95 per cent of the patients with 'osteogenesis imperfecta congenita' were said to have blue sclerae, but this assessment was probably made at birth, and it is not possible to say if they were re-examined in adult life.

Variable clinical picture

Study of any group with osteogenesis imperfecta shows that there are clinical differences between individuals, to which we have drawn attention. In 62 cases of osteogenesis imperfecta (Table 7.II) we were impressed by two unexpected findings which justify emphasis, namely that patients with fractures at birth did in fact survive to adolescence or adult life, and that in these patients the sclerae tended to be normal in colour. We could not estimate how many infants with osteogenesis imperfecta congenita had died early before we were able to see them and it is clear that the incidence of normal-coloured sclerae varies considerably according to the observer and patient population. It was also noted that patients within the severe group might differ widely, which suggested separate disorders. Differences are often based on x-ray appearances (p. 191), where it is difficult to

TABLE 7.II
Clinical findings of 62 patients with osteogenesis imperfecta
(From Smith, Francis and Bauze, 1975)

		Mild	Moderate	Severe	Total
Patients		25	4	33	62
Age in years:	average	25	16	20	21
	range	4–65	4–43	0–56	0–65
Sex:	Male	6	1	17	24
	Female	19	3	16	38
Family history		20	3	8	31
No. of families		12	3	30	44
Sclerae colour:					
blue		22	2	1	25
intermediate pale blue		3	1	11	15
normal		0	1	21	22
Fractures					
Number:	average	8	14	45	–
	range	1–25	7–30	5–100	–
Age of onset (years):	average	3½	1	1	
	range	0–18	1–1½	0–11	
Hyperplastic callus		0	1	4	5
Scoliosis		0	1	28	29
Unable to walk		1	0	21	22
Deafness		7	1	2	10
Dentinogenesis imperfecta		1	0	13	14

The age is at the time of examination. A positive family history indicates affected parents or siblings. Scoliosis was assessed by clinical examination.
For further details *see* Smith, Francis and Bauze (1975).

TABLE 7.III
Comparison of the clinical findings in two series of patients

Type of bone disease	Mild (no long-bone deformity)		Severe (long-bone deformity)	
Number of patients	25	35	33	55
Blue sclerae	25	27	12	49
Dentinogenesis imperfecta	1	20	13	39
Scoliosis	0	4	28	31
Deafness	7	4	2	8

For the purposes of this comparison, patients in the New York series (Falvo, Root and Bullough, 1974) labelled as osteogenesis imperfecta congenita and osteogenesis imperfecta tarda type 1 are included together under the heading of severe bone disease, since in both groups the incidence of long-bone deformity was 100 per cent. Oxford figures (Smith, Francis and Bauze, 1975) on the left, New York on the right of each group.
Note that patients with severe bone disease often have scoliosis

188

know whether some of the bizarre changes, especially in the lower limbs, are the result of prolonged disuse or orthopaedic treatment. Two interesting clinical sub-groups are those patients who form hyperplastic callus and those patients with severe disease in siblings.

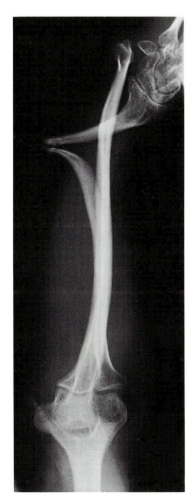

Figure 7.6. To demonstrate thin bones in a patient with severe osteogenesis imperfecta. The bizarre deformity follows a fracture

In the former patients masses of bony tissue, referred to as callus, are produced with or without obvious fracture. The affected limb becomes red and swollen and sarcoma may be simulated. Where biopsy has been done, the 'callus' may consist of fibromucoid and cartilage-like tissues. After the initial swelling, the bone slowly remodels, but bizarre changes persist (Bauze, Smith and Francis, 1975, Figure 11). The cause is quite unknown. Interestingly in two out of three families personally seen, it occurred in both parent and offspring, suggesting dominant inheritance. The patients of King and Bobechko (1971) with hyperplastic callus tended to be males with white sclerae.

Severe bone disease in more than one offspring of clinically unaffected parents is very rare and strongly suggests recessive inheritance producing an enzyme

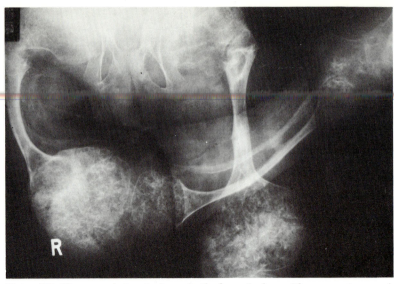

Figure 7.7. The expanded metaphyses in the legs of a boy with severe osteogenesis imperfecta who has never walked. The area of the growth plate is grossly disorganized, and has been replaced by many circular 'cystic' areas

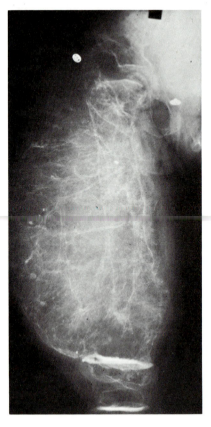

Figure 7.8. Examples of hyperplastic callus. in osteogenesis imperfecta. (a) In a girl in whom the whole femur appears to be replaced by a fine network of trabeculae

(a)

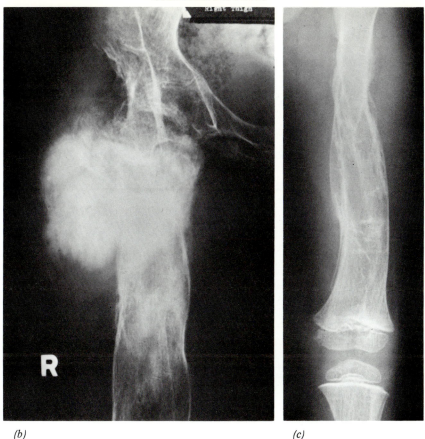

(b)   (c)

Figure 7.8 (b) Hyperplastic callus in a woman in whom the femur is deformed from previous episodes, and recent excessive callus has been produced. The radiograph of her daughter's femur (c) also shows the effects of hyperplastic callus

defect. Horan and Beighton (1975) describe two closely related families in whom 6 individuals had severe osteogenesis imperfecta, and in whom transmission was consistent with autosomal recessive inheritance. Of the 3 patients examined only 1 had definitely blue sclerae. Evidence for the existence of a recessive form of osteogenesis imperfecta is reviewed in detail by McKusick (1972).

Radiology

The x-ray appearances of osteogenesis imperfecta are very variable and often inexplicable, especially in those with severe bone disease. The appearances of some personal cases are shown (*Figures 7.6–7.11*). In the neonatal period, healing intra-uterine fractures and recent fractures associated with birth are seen. The long bones, such as the femur, may be short, wide and bent (*Figure 7.4*). This has been described as the 'thick bone' type (Fairbank, 1951). It is not clear whether the shortness of the limbs relative to the trunk at birth results entirely from the intra-uterine fractures or is a feature of generally disorganized growth. Rib fractures may be seen (*see* Figure 2 of Schoenfeld, Fried and Ehrenfeld, 1975) and are functionally important. The skull has poorly developed centres of ossification (wormian bones).

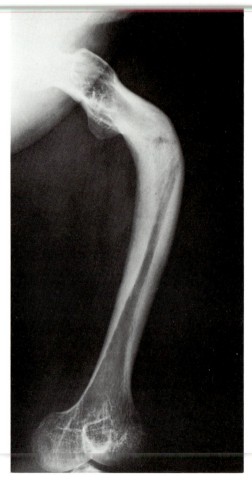

Figure 7.9. Buttressing of the femur in a woman with osteogenesis imperfecta and repeated fractures

After the neonatal period, many features may develop. The bones may become excessively thin (*Figure 7.6*); this thinness affects the shaft more than the ends of the bones which may appear to be expanded with cysts in them (*Figure 7.7*). Sometimes the whole shaft of the bone appears cystic. The end results of hyperplastic callus abolish the normal bony outlines (*Figure 7.8*). In some cases attempts at buttressing occurs (*Figure 7.9*). It is not clear whether the gross bending of bones sometimes seen occurs without fracture, but this certainly seems possible. Bowing of the lower one-third of the tibiae is a typical deformity, but the femora also may develop bowing (*Figure 7.10*).

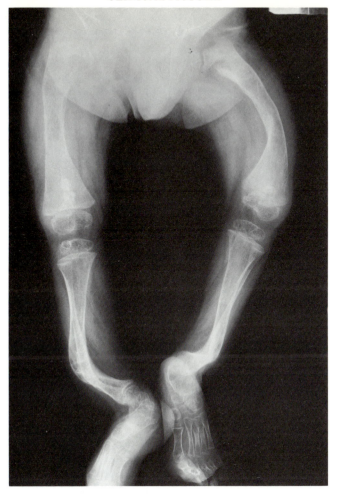

Figure 7.10. To demonstrate the typical tibial deformities in severe osteogenesis imperfecta. The femora are also abnormal and there is evidence of disruption of the growth plate with extension in to the metaphyses

Biochemistry

Routine biochemical measurements show little abnormality. Total urinary hydroxyproline (THP) excretion is probably normal (Lancaster *et al.*, 1975; Smith, Francis and Bauze, 1975), or reduced (Riley, Jowsey and Brown, 1973) for age. Since THP excretion reflects predominantly bone collagen turnover it normally changes with age and growth (Chapter 2), with a sharp temporary increase in adolescence, and interpretation of a single reading may be difficult. The validity of using a correction for surface area based on height and weight in severely deformed patients is doubtful, but it is equally fallacious to use age since the bone mass in severe osteogenesis imperfecta must be a very small fraction of normal. Thus for the size of the skeleton, the THP excretion must be disproportionately high.

Apart from such measurements which reflect collagen metabolism (and therefore might be expected to be abnormal), the large number of other recorded abnormalities are, on the whole, unconfirmed or inexplicable. These are an increase in basal metabolic rate and thyroxine levels (Cropp and Myers, 1972), alterations in leucocyte function (Humbert, Solomons and Ott, 1971) and changes in pyrophosphate metabolism. Since pyrophosphate may be important in normal calcification (Chapter 1), acting as an inhibitor, such changes (if true) would be very significant. Pyrophosphatases, which are normally responsible for pyrophosphate breakdown, are activated by magnesium ions. Solomons and Styner (1969) reported that both urine and serum pyrophosphate are increased in osteogenesis imperfecta but doubt has been cast on these results, partly because age-matched controls were not used and because plasma pyrophosphate was subsequently found to be normal (Russell *et al.*, 1971). Nevertheless the administration of magnesium oxide reduced elevated values (Solomons and Styner, 1969) but does not appear to alter bone morphology (Riley, Jowsey and Brown, 1973).

This area of research remains confused. Recently Granda, Falvo and Bullough (1977) found that the serum pyrophosphate was elevated in 19 out of 52 patients with osteogenesis imperfecta. There was no correlation between pyrophosphate and disease severity. Magnesium oxide or placebo did not affect the serum pyrophosphate or the clinical course.

Pathology

The pathology of osteogenesis imperfecta is also confusing and the confusion is made worse by the manner of its description and the apparent mixture of clinical types, of trabecular and cortical bone, and of woven and lamellar bone. The tissues most clearly involved are the bones, teeth and sclerae.

In mild osteogenesis imperfecta gross examination of the bone surface confirms osteoporosis. Electron microscopy supports the idea that the primary abnormality is in the osteoblast (Doty and Matthews, 1971). The excess of osteocytes is probably apparent rather than real because of the reduction in intercellular matrix. Thus reduction in distance between the osteocytes follows from the reduced width of the osteoid lamellae, which can be demonstrated under polarized light. In a complex article, Robichon and Germain (1968) emphasize the hyperosteocytosis and the apparently-increased bone resorption. It is not clear what types of osteogenesis imperfecta are being considered. Riley, Jowsey and Brown (1973) in a study of iliac crest biopsies of 11 patients with mild osteogenesis imperfecta found no abnormality in the ultrastructure of the osteocytes or bone collagen. Cortical thickness, osteoid width and number of osteons were reduced. Jowsey (1977) gives a good summary, emphasizing that bone biopsy findings indicate abnormal bone turnover. She points out that it is the presence of primary woven bone which is responsible for the large osteocyte lacunae, and that these have been misinterpreted as an indication of increased resorption. Osteoblasts are said to appear abnormal, being densely stained and fibroblastic in appearance.

194

The appearance of bone will alter considerably according to the clinical severity and type of disease.

In severe osteogenesis imperfecta (Follis, 1952) further difficulties in interpretation occur. The bone may be very immature and resemble foetal woven bone, upon which it is very difficult to make meaningful measurements. It seems that the gross arrangement of the bone matrix is abnormal, but there is little histological evidence of abnormality of the collagen fibres themselves (Falvo and Bullough, 1973). Scanning electron microscopy shows a disorganized pattern (Lindenfelser *et al.*, 1972; Teitelbaum *et al.*, 1974). There is little work on the teeth, but the major defect is one of dentine (Winter, 1969; Rao and Witkop, 1971). The teeth are excessively opalescent, the roots of the teeth are short and slender.

The available data show that in the mild form the collagen fibres in the skin are thinner than normal. The thickness of the sclerae may also be reduced and the diameter of the corneal collagen fibres diminished (Riley, Jowsey and Brown, 1973). Deafness, when present, appears to be due to fixation of the stapes by excessive bone formation.

Differential diagnosis

Diagnosis of osteogenesis imperfecta in the baby with intra-uterine fractures or fractures at birth, or in the older child with brittle bones and blue sclerae in a family similarly affected, should present no difficulty. However, there are times when the diagnosis is not clear and it should always be recalled that blueness of the sclerae is not a reliable physical sign of osteogenesis imperfecta in neonates.

At birth it is important to distinguish osteogenesis imperfecta from hypophosphatasia and from achondroplasia; later the main distinctions are from the battered baby syndrome and juvenile osteoporosis.

In severe osteogenesis imperfecta the short limbs of the baby simulate various varieties of short-limbed dwarfism, (McKusick, 1972) of which achondroplasia (Chapter 2) is the commonest. In severe hypophosphatasia (Chapter 11), bony deformity may occur, there is very poor mineralization of the skeleton, the alkaline phosphatase is low and there is an increased amount of phosphoethanolamine in the urine. In this disease antenatal diagnosis may be attempted.

In the infant with fractures it is important to distinguish between undiagnosed osteogenesis imperfecta and the 'battered baby' syndrome (Cameron and Rae, 1975). In the majority of infants with unexplained fractures the skeleton is normal and the injuries have been inflicted by parents or other adults. Such 'battered babies' often show features of ill-treatment and neglect with multiple bruising. Occasionally parents of infants with undiagnosed osteogenesis imperfecta may be accused of 'baby battering', and the possibility of osteogenesis imperfecta should always be considered.

In later childhood juvenile osteoporosis (Chapter 3) has to be distinguished. Again it is impossible to exclude mild osteogenesis imperfecta but the child with juvenile osteoporosis has certain important distinguishing features; onset is usually late in childhood or in early adolescence, in a previously normal skeleton; there is progressive but often self-limiting loss of height or delayed growth with vertebral collapse; partial metaphyseal fractures of the long bones may occur;

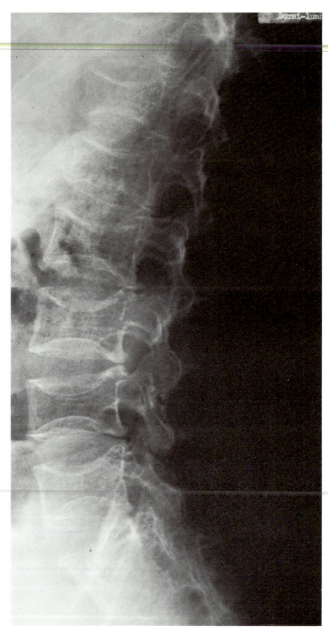

Figure 7.11. The spine in osteogenesis imperfecta. This is indistinguishable from the changes in juvenile osteoporosis. The vertebrae show evidence of multiple compression fractures, and each vertebra is biconcave

there is no family history and the sclerae are not obviously blue. The radiographic appearance of the spine in osteogenesis imperfecta may be indistinguishable from that of juvenile osteoporosis with multiple biconcave vertebrae which could be called a 'fish spine' (*Figure 7.11*).

TREATMENT

Logical treatment of osteogenesis imperfecta is not yet possible since we do not know its cause. However much can be done by a combination of orthopaedic surgery and rehabilitation especially in patients with severe bone disease. It is well to remember that although the severely deformed child may never walk despite surgery he is of normal intelligence and should be educated accordingly.

Orthopaedic treatment

In mild cases fractures heal well with conventional treatment, but difficulties often arise where there is severe bone deformity. The use of intramedullary rods, which may be extensible, to prevent further deformity and fracture and to give affected children sufficient strength to walk, has been widely discussed (King and Bobechko, 1971; Williams *et al.*, 1973) and can be very successful. Best results are obtained by treatment from birth in specialized centres. It may be difficult to prophesy which patients have bones strong enough to benefit from surgical treatment. The older child with grossly deformed lower limbs who has never walked is unlikely to do so whatever procedures are used. The use of inflatable 'space suits' is promising although experimental.

Rehabilitation

Severely affected children who cannot walk may spend their days in institutions or at home and rely on makeshift push-chairs for mobility. It is clearly important, where at all possible, that such patients should be assessed at rehabilitation centres and have properly constructed electrically-driven devices to allow them independent mobility (Goble and Nichols, 1971).

Medical treatment

In the child with severe osteogenesis imperfecta respiratory infections require prompt treatment. So far as the bone disease is concerned, it is important to protect the child from ill-advised treatments of no scientific worth, without discouraging justifiable therapeutic trials.

It is natural that attempts should have been made to strengthen the bones medically, either by correcting supposed deficiencies or by reducing supposed resorptive overactivity (Castells, 1973). However, there is no evidence that any alteration in diet, or increase of vitamins, or administration of anabolic steroids, or fluoride (Albright and Grunt, 1971), or calcium has any effect on the bone disease. The study of Kurz and Eyring (1974) demonstrates some of the difficulties in interpretation. They gave ascorbic acid for a variable length of time to patients of various ages with different types of osteogenesis imperfecta and noticed increased well-being, decreased fracture rate and lightening of the scleral colour. The first could be a placebo effect (no placebo trial was done) and the second two can occur with increasing age. Those who interpret the increase in osteocytes (and sometimes increased THP excretion) as due to excessively rapid bone turnover have used calcitonin in children (Castells *et al.*, 1972), and in adults (Caniggia and Gennari, 1972; Goldfield *et al.*, 1972). Although it is possible to reduce hydroxyproline excretion and to improve calcium balance, there is no convincing evidence of clinical improvement. In a recent study Rosenberg and his colleagues (1977) report the effects of salmon calcitonin for up to 3 years, given three times weekly to 10 patients with osteogenesis imperfecta. The biochemical results were variable and the fracture rate appeared to decrease in 3 patients. Although the authors suggested further studies, the results were quite inconclusive. It should be remembered that calcitonin has to be injected and may produce anorexia and vomiting, which limits its practical long-term use in small children (August, Shapiro and Hung, 1977).

Since magnesium is necessary for the function of pyrophosphatases, which may be reduced in osteogenesis imperfecta, patients have been treated with oral magnesium oxide. The results of this are not clear-cut, and despite early reports suggesting clinical improvement, it now seems that this treatment has little effect either on symptoms or biochemistry (Granda, Falvo and Bullough, 1977).

PROGNOSIS

Recent reviews of this disorder enable one to give a reasonably accurate prognosis. Babies of apparently normal parents, born with multiple fractures and the characteristic deformities of osteogenesis imperfecta 'congenita', may survive for many years, although early death from respiratory infection is common. In the present state of treatment they are unlikely to walk, but specialist orthopaedic advice should be obtained early. Children born into a family with a dominantly-inherited history of osteogenesis imperfecta have a far better prognosis, even if fractures occur at birth or soon after.

CAUSE

The cause of osteogenesis imperfecta continues to elude us. The clinical features suggest an abnormality of collagen, although a primary or additional abnormality of mucopolysaccharides (Dixon, Millar and Veis, 1975) is also possible. Since there are changes in connective tissue other than bone it seems unlikely that there is a primary disorder of calcification.

Until recently it has been difficult to demonstrate abnormal collagen in tissues or in the fibroblasts grown from them. Tissue analysis has concentrated on skin and shown that the amount of collagen in the skin may be reduced (Stevenson, Bottoms and Shuster, 1970), that this reduction is due to a reduction in the amount of the major polymeric collagen fraction of skin (Francis, Smith and Bauze, 1974), and that the ratio of Type III to Type I skin collagen is abnormally high for age, possibly due to a reduction in Type I collagen (Sykes, Francis and

Smith, 1977; *Figure 7.12*; Fujii, Kajiwari and Kurosu, 1977). These observations largely apply to mild osteogenesis imperfecta and suggest that there is a defective production of Type I collagen in skin. Since Type I collagen is the only type so far identified in bone, such a reduction could account for deficient formation of bone matrix in the mild dominantly-inherited disorder.

In severe osteogenesis imperfecta the situation is not so clear and the disorder may be heterogeneous. However, there is evidence that the polymeric collagen of skin is unstable (Smith, Francis and Bauze, 1975) but that its total amount may be normal. Likewise the ratio of Type I to Type III collagen in pepsin digests of skin may also be normal. The limited studies on clinically-unaffected relatives show no abnormality of this ratio.

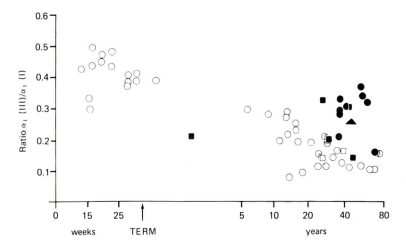

Figure 7.12. To demonstrate the abnormal ratio of the α chains of Type III collagen to Type I collagen in the skin of patients with mild osteogenesis imperfecta. Control subjects (○), mild osteogenesis imperfecta (●), severe osteogenesis imperfecta (■), clinically unaffected family members (□). In one case the disorder had features of both severe and mild groups (▲). (Reprinted by permission from Sykes, Francis and Smith, (1977) *New England Journal of Medicine*, **296**, 1200–1203)

Both Eastoe, Martens and Thomas (1973) and Trelstad, Rubin and Gross (1977) analysed tissues from osteogenesis imperfecta subjects and found an increase in lysyl hydroxylation which was most marked in the collagen of bone, and was associated with proportionate increases in covalently-bound glucose and galactose.

Since fibroblasts produce collagen they have been cultured from the skin of patients with osteogenesis imperfecta, and the synthesis of collagen examined. Some work suggests that the synthetic rate is normal (Lancaster *et al.*, 1975), others that it is diminished (Riley, Jowsey and Brown, 1973), and some results have been interpreted to show a selective reduction of Type I collagen (Penttinen *et al.*, 1975; Marcus and Bullough, 1977). However, fibroblasts in culture may not necessarily produce the same ratio of genetic collagens as in the tissue from which they were grown. Abnormal morphology of such fibroblasts has been reported (Lancaster *et al.*, 1975).

Discovery of the cause, or causes, of this disorder may take some time despite the considerable advances in biochemical knowledge, partly because tissue for analysis is necessarily limited. Some differences of opinion about the possible cause may be followed in the *Lancet* of 1976, referred to by Smith (1977). Recently Sykes and Solomon (1978) using the techniques of somatic-cell genetics have assigned a Type I collagen structural gene to human chromosome 7.

CURRENT PROBLEMS

Current problems are clinical and biochemical, and these are partly related. Most clinical problems arise from the management of severe osteogenesis imperfecta, as, for instance, the difficulty in predicting which is the appropriate case for attempted correction of deformity. Despite advances in biomechanics there seems no way of measuring the strength of small samples of bone, or predicting whether with the help of metal the long bones can withstand the weight of the body. Further clinical studies are important, again in the severe group of patients, to define how many disorders there are and what is their outcome. It is likely that more advance will come from regarding osteogenesis imperfecta as a group of separate disorders rather than as a single one. Biochemical definition of the underlying defect would help in management even if it would not suggest specific treatment. Finally, it would help if there was some agreement on pathology.

SUMMARY

Osteogenesis imperfecta is one of the most severe of the inherited crippling diseases. Current evidence suggests that it is due to an inborn error of collagen metabolism, but the disorder appears to be clinically and biochemically heterogeneous. Future investigation and treatment should take this heterogeneity into account.

REFERENCES

Albright, J. A. and Grant, J. A. (1971). Studies of patients with Osteogenesis Imperfecta. *J. Bone Jt Surg.* **53A**, 1415–1425

August, G. P., Shapiro, J. and Hung, W. (1977). Calcitonin therapy of children with osteogenesis imperfecta. *J. Pediat.* **91**, 1001–1005

Bauze, R. J., Smith, R. and Francis, M. J. O. (1975). A new look at Osteogenesis Imperfecta. A clinical, radiological and biochemical study of forty-two patients. *J. Bone Jt Surg.* **57B**, 2–12

Bergstrom, L. (1977). Osteogenesis Imperfecta: otologic and maxillofacial aspects. *The Laryngoscope* **87**, 1–42

Cameron, J. M. and Rae, L. J. (1975). *Atlas of the Battered Child Syndrome.* Edinburgh and London: Churchill Livingstone

Caniggia, A. and Gennari, C. (1972). Calcitonin treatment in Ekman-Lobstein disease. *Calcif. Tissue Res.* **9**, 243–244

Caniggia, A., Stuart, C. and Guideri, R. (1958). Fragilitas ossium hereditaria tarda. Ekman-Lobstein disease. *Acta med. scand.* Suppl. **340**, 1–172

Castells, S. (1973). New approaches to treatment of Osteogenesis Imperfecta. *Clin. Orthop. Related Res.* **93**, 239–249

REFERENCES

Castells, S., Inamdar, S., Baker, R. K. and Wallach, S. (1972). Effects of porcine calcitonin in Osteogenesis Imperfecta Tarda *J. Pediat.* **80**, 757–762

Cropp, G. R. A. and Myers, D. N. (1972). Physiological evidence of hyper-metabolism in Osteogenesis Imperfecta *Pediatrics*, **49**, 375–391

Dixon, I. R., Millar, E. A. and Veis, A. (1975). Evidence for abnormality of bone-matrix proteins in Osteogenesis Imperfecta. *Lancet* **2**, 586–587

Doty, B. S. and Matthews, R. S. (1971). Electron microscopic and histochemical investigation of Osteogenesis Imperfecta Tarda. *Clin. Orthop. Related Res.* **80**, 191–201

Eastoe, J. E., Martens, P. and Thomas, N. R. (1973). The amino-acid composition of human hard tissue collagens in Osteogenesis Imperfecta and Dentinogenesis Imperfecta. *Calcif. Tissue Res.* **12**, 91–100

Eddowes, A. (1900). Dark sclerotics and fragilitas ossium. *Br. med. J.* **2**, 222

Ekman, O. J. (1788) *See* McKusick, V. A. (1972). p. 444

Fairbank, T. (1951). *An Atlas of General Affections of the Skeleton.* 1st edn. Edinburgh and London: E. and S. Livingstone

Falvo, K. A. and Bullough, P. G. (1973). Osteogenesis Imperfecta: a histometric analysis. *J. Bone Jt Surg.* **55A**, 275–286

Falvo, K. A., Root, L. and Bullough, P. G. (1974). Osteogenesis Imperfecta: clinical evaluation and management. *J. Bone Jt Surg.* **56A**, 783–793

Follis, R. H. (1952). Osteogenesis Imperfecta congenita: A connective tissue diathesis. *J. Pediat.* **41**, 713–720

Francis, M. J. O., Smith, R. and Bauze, R. J. (1974). Instability of polymeric skin collagen in Osteogenesis Imperfecta. *Br. med. J.* **1**, 421–424

Fujii, K., Kajiwari, T. and Kurosu, H. (1977). Osteogenesis Imperfecta: Altered content of Type III collagen and proportion of the cross-links in skin. *Fed. Europ. Biochem. Soc. Letters* **82**, 251–254

Goble, R. E. A. and Nichols, P. J. R. (1971). *Rehabilitation of the Severely Disabled.* London: Butterworths

Goldfield, E. B., Braiker, B. M., Prendergast, J. and Kolb, F. O. (1972). Synthetic salmon calcitonin. Treatment of Paget's disease and Osteogenesis Imperfecta. *J. Am. med. Ass.* **221**, 1127–1129

Granda, J. L., Falvo, K. A. and Bullough, P. G. (1977). Pyrophosphate levels and magnesium oxide therapy in Osteogenesis Imperfecta. *Clin. Orthop. Related Res.* **126**, 228–231

Gray, P. H. K. (1969). A case of Osteogenesis Imperfecta, associated with dentinogenesis imperfecta, dating from antiquity. *Clin. Radiol.* **20**, 106–108

Horan, F. and Beighton, P. (1975). Autosomal recessive inheritance of Osteogenesis Imperfecta. *Clin. Genetics* **8**, 107–111

Humbert, J. R., Solomons, C. C. and Ott, J. E. (1971). Increased oxidative metabolism by leukocytes of patients with Osteogenesis Imperfecta and of their relatives. *J. Pediat.* **78**, 648–653

Ibsen, K. H. (1967). Distinct varieties of Osteogenesis Imperfecta. *Clin. Orthop. Related Res.* **50**, 279–290

Jowsey, J. (1977). *Metabolic Diseases of Bone.* Vol. 1. London: W. B. Saunders Co

King, J. D. and Bobechko, W. P. (1971). Osteogenesis Imperfecta. An orthopaedic description and surgical review. *J. Bone Jt Surg.* **53B**, 72–89

Kurz, D. and Eyring, E. J. (1974). Effects of vitamin C on Osteogenesis Imperfecta. *Pediatrics*, **54**, 56–61

Langness, U. and Behnke, H. (1971). Collagen metabolites in plasma and urine in Osteogenesis Imperfecta. *Metabolism* **20**, 456–463

Lancaster, G., Goldman, H., Scriver, C. R., Gold, R. J. M. and Wong, I. (1975). Dominantly-inherited Osteogenesis Imperfecta in man; an examination of collagen biosynthesis. *Pediat. Res.* **9**, 83–88

Levin, L. S. and Thompson, R. G. (1975). Osteogenesis Imperfecta tarda presenting with short stature. In *Disorders of Connective Tissue.* D. Bergsma (Ed.), *Birth Defects Original Article Series* **XI**, 6, pp. 103–105

Levin, L. S., Salinas, C. F. and Jorgenson, R. J. (1978). Classification of osteogenesis imperfecta by dental characteristics. *Lancet* **1**, 332

Lindenfelser, R., Hasselkus, P., Haubert, P. and Kronert, W. (1972). Zur Osteogenesis Imperfecta congenita. Rasterelektronenmikroskopische untersuchungen. *Virchows Arch. Abt. B. Zellpath.* **11**, 80–89

REFERENCES

Marcus, R. E. and Bullough, P. G. (1977). Abnormal collagen synthesis by cultured dermal fibroblasts in Osteogenesis Imperfecta. *Arthritis and Rheumatism,* **20,** 127

McKusick, V. A. (1972). *Heritable Disorders of Connective Tissue.* 4th edn St. Louis: C. V. Mosby Co

Nordin, B. E. C. (1973). *Metabolic Bone and Stone Disease.* 1st edn. Edinburgh and London: Churchill Livingstone

Penttinen, R. P. Lichtenstein, J. R., Martin, G. R. and McKusick, V. A. (1975). Abnormal collagen metabolism in cultured cells in Osteogenesis Imperfecta. *Proc. natn. Acad. Sci. U.S.A.* **72,** 586−589

Rao, S. and Witkop, C. J. (1971). Inherited defects in tooth structure. The Third Conference on the Clinical Delineation of Birth Defects. *Birth defects. Original Article Series, XI.* **7,** 153−184

Riley, F. C., Jowsey, J. and Brown, D. M. (1973). Osteogenesis Imperfecta: morphologic and biochemical studies of connective tissue. *Pediat. Res.* **9,** 757−768

Robichon, J. and Germain, J. P. (1968). Pathogenesis of Osteogenesis Imperfecta. *Can. med. Ass. J.* **99,** 975−979

Rosenberg, E., Lang, R., Boisseau, J., Rojanasathit, S. and Avioli, L. V. (1977). Effect of long-term calcitonin therapy on the clinical course of Osteogenesis Imperfecta. *J. Clin. Endocr. Metabolism* **44,** 346−353

Ruedemann, A. D. (1953). Osteogenesis Imperfecta congenita and blue sclerotics: a clinico-pathological study. *Archs Ophthal.* **49,** 6−16

Russell, R. G. G., Bisaz, S., Donath, A., Morgan, D. B. and Fleisch, H. (1971). Inorganic pyrophosphate in plasma in normal persons and in patients with hypophosphatasia, Osteogenesis Imperfecta, and other disorders of bone. *J. clin. Invest.* **50,** 961−969

Schoenfeld, Y., Fried, A. and Ehrenfeld, N. E. (1975). Osteogenesis Imperfecta. Review of the literature with presentation of 29 cases. *Am. J. Dis. Child.* **129,** 679−687

Smith, R. (1977). Paget's disease of bone, Osteogenesis Imperfecta and fibrous dysplasia. *Br. med. J.* **1,** 365−367

Smith, R., Francis, M. J. O. and Bauze, R. J. (1975). Osteogenesis Imperfecta. A clinical and biochemical study of a generalized connective tissue disorder. *Q. Jl Med.* **44,** 555−573

Solomons, C. C. and Styner, J. (1969). Osteogenesis Imperfecta. Effect of magnesium administration on pyrophosphate metabolism. *Calcif. Tissue Res.* **3,** 318−326

Stevenson, C. J., Bottoms, E. and Shuster, S. (1970). Skin collagen in Osteogenesis Imperfecta. *Lancet* **1,** 860−861

Suen, V. F., Harris, V. and Berman, J. L. (1974). Osteogenesis Imperfecta congenita. Report of a mother and son. *Clin. Genetics* **5,** 307−311

Sykes, B. and Solomon, E. (1978). Assignment of a type I collagen structural gene to human chromosome 7. *Nature, Lond.* **272,** 548−549

Sykes, B., Francis, M. J. O. and Smith, R. (1977). Altered relation of two collagen types in Osteogenesis Imperfecta. *New Engl. J. Med.* **296,** 1200−1203

Teitelbaum, S. L., Kraft, W. J., Lang, R. and Avioli, L. V. (1974). Bone collagen aggregation abnormalities in Osteogenesis Imperfecta. *Calcif. Tissue Res.* **17,** 75−79

Trelstad, R. L., Rubin, D. F. and Gross, J. (1977). Osteogenesis Imperfecta Congenita: evidence for a generalized molecular disorder of collagen. *J. Lab. Invest.* **36,** 501−508

Williams, P. F., Cole, W. H. J., Bailey, R. W., Dubow, H. I., Solomon, C. C. and Millar, E. A. (1973). Current aspects of the surgical treatment of Osteogenesis Imperfecta. *Clin. Orthop. Related Res.* **96,** 288−298

Winter, G. B. (1969). Hereditary and idiopathic anomalies of tooth number, structure and form. *Dent. Clin. N. Am.* **13,** 355−373

Wynne-Davies, R. (1973). *Heritable Disorders in Orthopaedic Practice.* p. 80. Oxford and London: Blackwell Scientific Publications

8

Marfan's Syndrome, Homocystinuria, The Ehlers-Danlos Syndrome

INTRODUCTION

There are several inherited disorders of connective tissue in which, unlike osteo-genesis imperfecta, the skeletal manifestations are only one aspect of a complex clinical picture. Most of these probably involve collagen and of them Marfan's syndrome appears to be the most common (Table 8.I). They all deserve consideration because of the clinical problems which they produce and because of recent advances in knowledge of their biochemistry. Homocystinuria closely resembles Marfan's syndrome, for which it may easily be mistaken until the urine is tested for homocystine (Schimke *et al.*, 1965). Scoliosis is a common feature of Marfan's syndrome, of homocystinuria, and of many inherited disorders of connective tissue, and there are reasons to include so-called idiopathic scoliosis in the present chapter. Separate consideration is given to the condition known as congenital contractural arachnodactyly which is closely related to Marfan's syndrome. The classification of the several distinct disorders which together comprise the Ehlers–Danlos syndrome is still incomplete and depends both on the clinical picture and on the underlying biochemical abnormality, where this has been identified. Full descriptions of these disorders may particularly be found in the writings of McKusick (1972, 1976). For this chapter it is necessary to recall some points about collagen synthesis and chemistry (Chapter 1). The probable abnormalities in collagen in the disorders discussed in this chapter are shown in *Figure 8.1*. It is appropriate also to consider in this chapter two inherited disorders, one classic (alkaptonuria) and one new (Menkes syndrome), which indirectly affect connective tissues and in this respect come into the same category as homocystinuria (McKusick, 1976). The mucopolysaccharidoses are dealt with separately (Chapter 9).

TABLE 8.I
Inherited disorders of connective tissue with biochemical defects, identified or considered likely (excluding osteogenesis imperfecta)

	Disorder	Defect
Primary		
	Marfan's syndrome	Not identified
	Ehlers–Danlos (E–D) syndrome	Identified in Type IV, VI and VII (see Table 8.III)
Secondary		
	Homocystinuria	Cystathionine synthase deficiency
	Alkaptonuria	Homogentisic acid oxidase deficiency
	Menkes syndrome	Copper deficiency

MARFAN'S SYNDROME

Marfan's syndrome is a dominantly-inherited disorder characterized by skeletal deformity, arachnodactyly, dislocated lenses and aortic dilatation.

Pathophysiology

Since the exact biochemical defect in Marfan's syndrome can only be surmised, it is not possible to explain the pathological findings in such terms. Nevertheless there is an increasing understanding of the major abnormalities produced by this connective-tissue disorder, particularly in the cardiovascular system. These affect the aorta and the mitral (and tricuspid) valves and are the main cause of death.

In the aorta a weakness of the tunica media leads to progressive dilatation at its base, with marked aortic incompetence, and subsequently dissecting aneurysm. Dilatation of the pulmonary artery also occurs, but less commonly. Recent work has emphasized the importance of changes in the mitral valve which lead to regurgitation of a particular type. The valves themselves may be abnormal and the chordae tendineae redundant. The valves therefore tend to protrude into the left atrium in late systole.

The appearance of the tissues suggest an abnormality of collagen and/or elastin in this disorder, but the evidence that collagen is abnormal is not convincing. Early measurements on the excretion of hydroxyproline in the urine (Sjoerdsma et al., 1958) showed an increase but this was slight and was found in young patients at a time when the hydroxyproline might not have fallen to the normal adult levels. The collagen produced by fibroblasts from patients with Marfan's syndrome appears to be unusually soluble (Priest, Moinuddin and Priest, 1973), but the stability and amount of the polymeric collagen from skin was found to be normal (Francis, Sanderson and Smith, 1976). Certainly it has not been possible to demonstrate the well-defined collagen abnormalities which are a feature of homocystinuria.

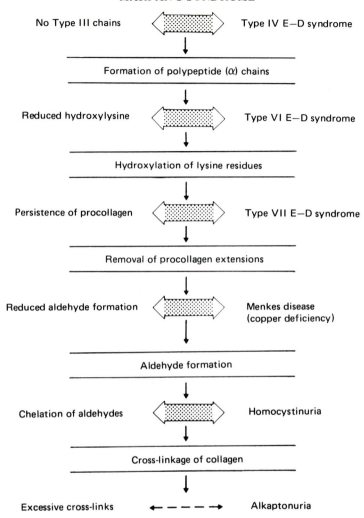

No Type III chains   Type IV E–D syndrome

Formation of polypeptide (α) chains

Reduced hydroxylysine   Type VI E–D syndrome

Hydroxylation of lysine residues

Persistence of procollagen   Type VII E–D syndrome

Removal of procollagen extensions

Reduced aldehyde formation   Menkes disease (copper deficiency)

Aldehyde formation

Chelation of aldehydes   Homocystinuria

Cross-linkage of collagen

Excessive cross-links   Alkaptonuria

Figure 8.1. Postulated or identified defects of collagen synthesis in inherited disorders of connective tissue. In this illustration and in Figures 8.3 and 8.7 stippled arrows indicate metabolic blocks

Clinical features

The expression of the clinical features of Marfan's syndrome is variable. Its features are skeletal, vascular and ocular; when all are present they make the characteristic triad of long thin extremities, aortic aneurysm and ectopia lentis.

It is clear that clinical heterogeneity exists. Thus McKusick (1972) divides Marfan's syndrome into at least four recognizable types, namely an asthenic form, a non-asthenic form, the Marfanoid hypermobility syndrome, and contractural arachnodactyly. It is certainly impressive that some patients with Marfan's syndrome are very thin, loose-jointed and scoliotic and appear to have very little subcutaneous tissue and muscle whilst others do not have these features. Contractures may occur in patients with the typical features of Marfan's syndrome; an example of this is shown in *Figure 8.2* and also in Figure 28 of Brenton *et al*. (1972). There is also a dominantly inherited condition which has been recognized separately and termed congenital contractural arachnodactyly.

Skeletal

Typically the patient with Marfan's syndrome is very tall and often (but not always) thin. The fingers are long and thin (arachnodactyly), the chest is often deformed with a protuberant and distorted sternum or with pectus excavatum;

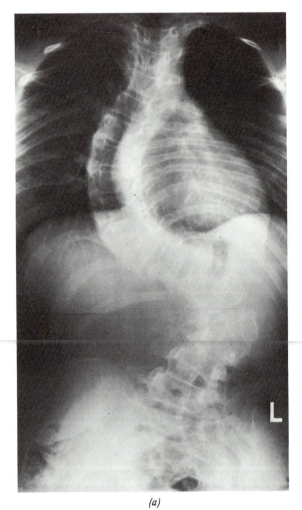

(a)

Figure 8.2. Marfan's syndrome in a girl of 11 years. (a)
Severe scoliosis

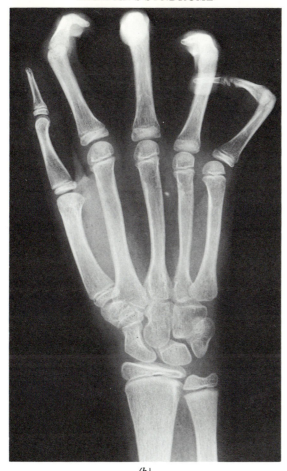

(b)

Figure 8.2. (b) Arachnodactyly with contractures. The lenses were dislocated and the urine did not contain homocystine

the limbs are disproportionately long compared with the trunk and the palate is high-arched. In measuring this disproportion it is necessary to recall the normal proportions of the body, their alterations with age and race, and the wide variation in the normal population (Chapter 2). Thus in Caucasian adults the upper border of the symphysis pubis is at approximately the mid-point of the body so that the ratio of upper to lower segment is approximately unity. McKusick (1972) finds that the mean adult upper segment/lower segment (US/LS) ratio in whites is about 0.93, with a range (± 2SD) of 0.85 to 1.0 (taken from his Figure 3-11). Brenton (1977) points out that McKusick derives the US/LS ratio from measurement of standing height and pubis to heel, with crown to pubis obtained by subtraction. Since trunk height is less on standing than lying, this will reduce the US/LS ratio. In negroes with relatively longer limbs, the US/LS ratio is lower than in Caucasians.

In classic cases of Marfan's syndrome, the US/LS ratio is much reduced and the obvious disproportion may be exaggerated by a coexistent scoliosis. In patients with mild Marfan's syndrome it is impossible to make the diagnosis from the skeletal features alone. The combination of long thin extremities and loose-jointedness forms the basis of many of the physical 'tests' for Marfan's syndrome. Thus the flexed thumb can extend across and beyond the ulnar margin of the hand, and the thumb and fifth finger of one hand overlap when the opposite wrist is grasped. Many other skeletal features are described (McKusick, 1972). One should be cautious in using an apparently high-arched palate in support of the diagnosis of Marfan's syndrome, since the shape of a palate may be altered by many factors. Scoliosis requires particular mention (p. 217). The described excessive length of the big toes is another manifestation of the skeletal disproportion. The weakness of the joint capsules and ligaments, tendons and fascia may lead to such features as dislocation of the joints and hernia.

Vascular

The main vascular change in Marfan's syndrome is the dilatation of the aorta, with its eventual rupture or dissection. The experience of this has been recorded by McKusick (1972) who points out that aortic dilatation may begin at the root of the aorta, within the cardiac silhouette, and that considerable aortic regurgitation may be present before dilatation is detectable on x-ray. Dissection of the aorta is most commonly limited to the ascending part. Valvular abnormalities, of which the commonest is a redundant or 'floppy' mitral valve, are being increasingly recognized. Severe mitral regurgitation may occur.

Ocular

Patients with Marfan's syndrome tend to have dislocation of the lenses, the cause of which is unknown. It is said that the lenses tend to dislocate upwards (however *see Figure 2.6*), in contrast to homocystinuria where the opposite may be found (*Figure 8.5*). When the lens is dislocated the red reflex of the retina will be seen above or below the margin of the lens, and the unsupported iris will shimmer when the eyes are moved (a phenomenon which is called iridodonesis). In Marfan's syndrome the lenses are usually dislocated (to some extent) at birth.

Diagnosis

The diagnosis of Marfan's syndrome is normally suggested by the skeletal abnormalities and confirmed by the finding of dislocated lenses and other features such as aortic incompetence or a prolapsing mitral valve. In the typical case a parent is also affected, being disproportionately tall and having dislocated lenses, and other family members may show the same features. In the doubtful case there is no certain way of excluding Marfan's syndrome by clinical or biochemical examination. It is important when considering the diagnosis in an excessively tall person to take into account the height of the parents, from which the height of the offspring may (within limits) be predicted.

MARFAN'S SYNDROME

TABLE 8.II

Comparison of the features of Marfan's syndrome (MS), homocystinuria (H)
and congenital contractural arachnodactyly (CCA)

	MS	H	CCA
Dislocated lens	+	+	−
Mental backwardness	−	+ variable	−
Aortic dilatation	+	−	−
'Floppy' mitral valve	+	−	−
Thrombosis (venous and arterial)	−	+	−
Arachnodactyly	+	+	+
Scoliosis	+	+	+
Osteoporosis	−	+	−
Contractures	Rare	−	Always
Inheritance	D	R	D
Homocystine in urine	−	+	−
Hydroxyproline in urine	Sometimes increased		
Fair hair, malar flush	−	+	−
Other	−	−	Abnormal helix of ear

D = dominant   R = recessive   N = normal

There are two steps in the diagnosis of Marfan's syndrome. The first is to decide which of the clinical features are present; and the second is to exclude other important conditions which simulate it. Thus disproportionately long limbs may occur in hypogonadism, in Klinefelter's syndrome (XXY chromosomal abnormality), in sickle-cell anaemia and in some negro races. Scoliosis is often 'idiopathic' in the adolescent and does not necessarily suggest Marfan's syndrome, and dominantly-inherited neurofibromatosis may cause scoliosis and disproportion. Dislocation of the lens often occurs in isolation. Important conditions which simulate Marfan's syndrome include homocystinuria (Table 8.II) congenital contractural arachnodactyly and, very rarely, medullary carcinoma of the thyroid (Chapter 1), in which a Marfanoid habitus is associated with mucosal neuromata.

Treatment

The main cause of death in Marfan's syndrome is aortic valve disease, and attempted prevention and treatment is very important. Other features which may require treatment include scoliosis and excessive growth, especially in girls. In a survey of 257 patients with Marfan's syndrome (Murdoch *et al.*, 1972), cardiac problems accounted for 52 out of 56 deaths of known cause and 39 of these were due to aortic dilatation and its complications. The cumulative probability of survival fell to 0.5 in men between 40 and 41 years of age; and the mean age of death for all 74 patients who died in this group was 32 years.

The characteristics of the pulse wave in Marfan's syndrome appear to be important in the extension and eventual rupture of dissecting aneurysms, and they may be modified by reserpine or propanolol with a reduction in the rate of aortic dilatation. Where necessary, surgical replacement of the ascending aorta and valve is increasingly successful.

Excessive rate of growth and eventual height may be very troublesome in girls. McKusick (1976) states that he uses a combined oestrogen–progesterone regime to initiate puberty early in such girls, to control the eventual height and to carry the girl through the period of the maximal risk of scoliosis. Since it is very difficult to predict eventual height accurately, such hormonal manipulations are worth considering only in severe cases.

Of the other skeletal abnormalities which require surgical correction, scoliosis is the most common. Winter (1977) finds that there are close clinical and surgical similarities to the 'idiopathic' group.

HOMOCYSTINURIA

Biochemical defect

This rare disorder is due to a deficiency of the enzyme cystathionine synthetase (or synthase) (*Figure 8.3*). There is a block in the conversion of homocysteine to cystathionine. Homocysteine accumulates proximal to the block and is

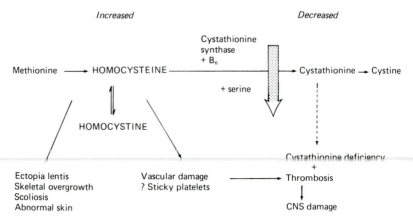

Figure 8.3. The enzyme block in homocystinuria and its possible effects

converted to homocystine, which is present in increased amounts in the plasma and urine, where it may be detected by the nitroprusside test. There is also an increase in plasma methionine concentration. Cystathionine, normally present in the brain, is no longer detectable. Cystine cannot be made from methionine and becomes an essential amino acid in the homocystinuric patient. Various causes of homocystinuria exist but this description will confine itself to the 'classic' inherited form in which the enzyme block may or may not be overcome by large doses of pyridoxine (vitamin B_6). Mudd and Levy (1978) provide a detailed discussion of the biochemistry. Vitamin B_6-responsive homocystinuria is one of the vitamin-dependent disorders, like vitamin-D-dependent rickets (p. 127).

In the years following its first discovery (Field et al., 1962; Carson and Neill, 1962) it was found that the biochemical abnormalities in this disorder could sometimes be corrected by large doses of vitamin B_6 (for review see Cusworth and Dent, 1969). Subsequently the enzyme defect, which could be demonstrated in liver, brain, skin fibroblasts and phytohaemagglutinin-stimulated lymphocytes, was found to be very variable (Gaull, Sturman and Schaffner, 1974). The biochemical response to pyridoxine depended on the amount of enzyme present, and the B_6-unresponsive patients had the lowest enzyme levels and the most striking clinical features. Heterozygotes, who are clinically normal, had reduced levels of cystathionine synthase in their tissues and fibroblasts and could also be detected by delayed metabolism of a standard oral-methionine load (Sardharwalla et al., 1974). It should be possible to make an antenatal diagnosis of homocystinuria from the cystathionine synthase content of amniotic fluid cells (Fleisher et al., 1974).

Pathophysiology

In homocystinuria there are widespread vascular lesions, with dislocation of the lenses, skeletal abnormalities and a variable degree of mental deterioration. It is not possible to explain all the known pathological changes in terms of the bio-chemical ones but some suggestions may be made. Theoretically the clinical features could be due either to accumulation of substances such as methionine or homocystine proximal to the metabolic block, or to a deficiency of substances such as cystine and cystathionine distal to it. Although the cause of the vascular lesions, which include intimal fibrosis, abnormal elastic laminae, atherosclerosis and thrombosis, is still debated, it seems likely to be related to accumulation of homocysteine. Some authors have stressed the importance of platelet stickiness, whilst others (Harker et al., 1974) have demonstrated endothelial lesions in the blood vessels of primates exposed to artificially high concentrations of homocystine. Whatever the cause, which remains in doubt (Uhlemann et al., 1976), there is certainly a tendency towards thrombosis, which may account for the progressive mental deterioration (see Brenton et al., 1966). It is not known why the lenses dislocate, and this is unlikely to be discovered until the normal chemistry of the suspensory ligament of the lens is elucidated (see Mudd and Levy, 1978). The skeletal abnormalities, which are closely similar to those of Marfan's syndrome, also remain unexplained but the scoliosis may in part be due to the collagen abnormality which the homocyst(e)ine appears to produce; and the abnormal growth may possibly be related to the somatomedin-like effect of homocystine (Dehnel and Francis, 1972). The increase in concentration of homocysteine and homocystine produced by the partial enzyme block interferes with the cross-linking of collagen by chelating with the aldehydes derived from lysine and hydroxylysine. It has been suggested that the biochemical effects of homocysteine are similar to those of penicillamine, used in some patients with rheumatoid arthritis and in Wilson's disease. Francis, Smith and MacMillan (1973) showed that the stability of cross-linked skin collagen to depolymerization from patients with homocystinuria was reduced compared with age-matched controls (Chapter 1); and Kang and Trelstad (1973) showed that neutral salt solutions of purified rat-skin collagen failed to form insoluble fibrils when exposed to homocysteine, whereas homocystine or methionine lacked this effect, and that dermal collagen from homocystinuric patients was significantly more soluble in non-denaturing solvents and had less cross-links than normal. Subsequently Griffiths, Tudball and Thomas (1976) found a variable response in rats to intraperitoneal homocystine and methionine. In some animals who died there was evidence of deficient cross-links in tendon collagen and aorta elastin, whereas in those who survived no such defect was found. This variability was postulated to be due to variation in the circulating levels of homocysteine, itself a very toxic substance. One should recall that since the cross-linking of elastin depends on similar mechanisms to collagen, it is likely (although not demonstrated), that elastin is also abnormal in homocystinuria.

Clinical features

Homocystinuria is rare and recessively inherited. Thus it occurs in siblings but not through generations (*Figure 8.4*). Its main features are mental backwardness, dislocation of the lens, skeletal abnormalities closely resembling Marfan's

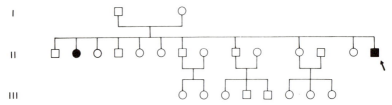

Figure 8.4. Pedigree of a patient with pyridoxine-sensitive homocystinuria, who was the youngest child of unrelated parents. His sister who was almost certainly affected (although the urine was never tested for homocystine) died in her early 20s in a mental hospital following a 'thrombosis'. Both she and her affected brother were fair-haired − the rest of the family had dark hair. Propositus ■; Probable affected sibling ●

syndrome; and a tendency to venous and arterial thrombosis. Other described features include a fatty liver, fair hair and a malar flush. Good descriptions of the clinical features are given by Schimke *et al.*, (1965), Brenton *et al.*, (1966) and Carey *et al.*, (1968). The latter noted particularly 'tissue paper' scars of the skin. Those who are pyridoxine-responsive tend to be less severe than those who are not and the former may be of normal intelligence (McKusick, 1976).

Mental backwardness

The degree of mental defect in homocystinuria is very variable and in some series nearly half have normal intelligence but it has not been proved that the mental state is altered by treatment with pyridoxine despite correction of the biochemical disturbance. Likewise we do not know whether the mental retardation is due to intermittent cerebral thrombosis or to biochemical abnormalities, but it is less marked in B_6-responders. Other central nervous system disorders may occur; in early childhood 'acute hemiplegia' and 'cerebral palsy' may result from intracranial thrombosis, and carotid artery thromboses are recorded. Schizophrenia-like states have been described.

Dislocation of the lens

This is common (*Figure 8.5*). A good example is also illustrated by Mudd and Levy (1978). It is said that in contrast to Marfan's syndrome the dislocation is not present at birth. Such dislocation often produces glaucoma.

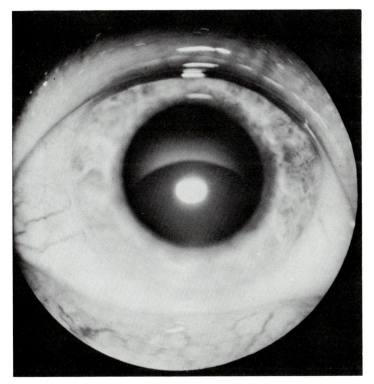

Figure 8.5. Dislocated lens of propositus in Figure 8.4. The lens had dislocated downwards. In colour the lens appeared blue and the crescentic shape area above it red

Skeletal changes

The patient with homocystinuria may closely resemble the patient with Marfan's syndrome sometimes with disproportionately long limbs and arachnodactyly Brenton (1977). Joint mobility, however, is said to be reduced rather than increased. There are some additional features. In some patients the hands are very large, and the bones as a whole are very big; this includes the vertebrae, and

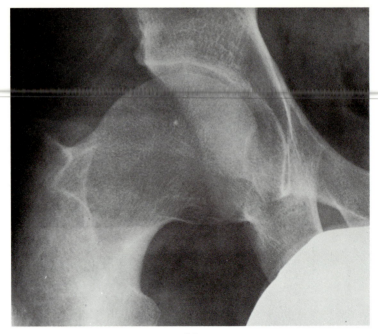

(a)

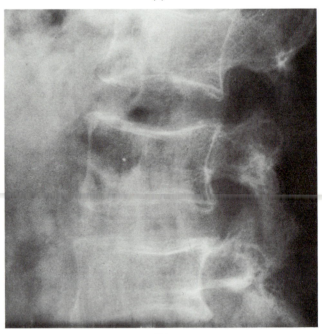

(b)

Figure 8.6. Appearance of radiographs of propositus in Figure 8.4.
(a) Excessively large femoral head. (b) Radiolucent vertebrae with
posterior biconcavity

214

the femoral heads. Scoliosis is common, but contractures, which sometimes occur in Marfan's syndrome, are not present (Table 8.II). The skeletal features have been well reviewed by Brenton and his colleagues (1966, 1972), who found that widening of the epiphyses and metaphyses of the long bones with flattening or posterior biconcavity of the vertebrae were common features. Multiple Harris lines were common; the femoral head sometimes appeared large and the femoral neck abnormal in shape. Patients with homocystinuria are also said to have 'osteoporosis'. This is most often inferred from the radiological appearances. McKusick (1972) describes other features, with enlarged malformed carpal bones. In one of the early cases from Oxford, Stuart M. (Brenton *et al.,* 1966), the similarity to acromegalic gigantism led to the measurement of growth hormone which was found to be normal. The skeletal deformities appear to be progressive and are not present at birth. Brill, Mitty and Gaull (1974) concluded that the skeletal changes were most severe in pyridoxine-non-responsive homo-cystinuria with the severest deficiency of cystathionine synthase. They remarked on the unusual appearance of calcific spicules in the radius or ulna extending from the growth plate in children into the shaft. Some radiological features in the patient with pyridoxine-sensitive homocystinuria, whose pedigree appears in *Figure 8.4,* are shown in *Figure 8.6.*

Vascular complications

Venous and arterial thromboses are major important features of homocystinuria. They occur particularly after surgical procedures, so that any form of elective surgery should be avoided if possible. Spontaneous thromboses can occlude the coronary, renal and cerebral arteries, and peripheral venous thromboses may lead to pulmonary embolism.

Other features

Many patients with homocystinuria may have a typical facial appearance with malar flushing and light-coloured hair which darkens on pyridoxine therapy. The skin is often thin and marbled in appearance. The teeth are crowded and herniae may occur. However in mild cases the appearance may be virtually normal.

Diagnosis

The important differential diagnosis is from Marfan's syndrome (Table 8.II). In every patient with the features of Marfan's syndrome the urine should be tested for homocystine by the nitroprusside test, together with a normal control. Homocystine in the urine is first reduced to homocysteine by sodium cyanide. Homocysteine reacts with freshly prepared sodium nitroprusside to form a dark-red colour. If this test is positive in a patient with 'Marfan's syndrome' he is likely to have classic homocystinuria. Cystine gives the same reaction in the nitroprusside test, but may be distinguished from it by moving at a different rate on high-voltage electrophoresis (McKusick, 1972, Figures 4—11). The chances that a patient with the physical features of Marfan's syndrome and a positive nitroprusside test has cystinuria rather than homocystinuria must be excessively low.

Treatment

Once the diagnosis of 'classic' homocystinuria is made it is logical to attempt to reverse the biochemical abnormalities with pyridoxine, which is given initially in large doses, up to 250 mg four times daily, and subsequently reduced to about 50 mg daily. In pyridoxine-sensitive homocystinuria this treatment will overcome the metabolic block and correct the amino acid abnormalities. Homocystine will then no longer be detectable in the urine. It is not clear whether correction of the biochemical defect with pyridoxine will also improve the clinical state or prevent its progression, but it seems reasonable to continue such treatment. McKusick (1976) states that pyridoxine treatment of B_6-responsive homocystinurics can prevent thrombotic episodes and even ectopia lentis. Brenton et al., (1977) describe normal pregnancies and offspring in women with pyridoxine-responsive homocystinuria. Patients with homocystinuria have skeletal and ocular defects which may require operative correction or they may need surgery for unrelated reasons. Postoperative thrombosis is common but may be reduced by giving aspirin 300 mg twice daily.

Finally, one should note that homocystinuric patients may have low levels of serum folate, possibly due to its increased utilization, and that folic acid should be given as an adjunct to pyridoxine (Wilcken and Turner, 1973).

Discussion

In homocystinuria the association of a measurable and potentially reversible biochemical defect with well-defined clinical features has presented many interesting problems. The basic problem has been to relate the high circulating levels of homocystine to the clinical manifestations. Although there is an increase in circulating homocystine, it is likely that the deleterious effects are due to the monosulphide homocysteine. The abnormal collagen cross-linking and the somatomedin-like effect of homocystine may account for the abnormal skeletal appearance, but there are difficulties in explaining the thrombotic tendency. Thus Harker et al., (1974) showed that in 4 patients with homocystinuria, platelet survival was uniformly shortened whilst platelet function tests were normal. In 2 pyridoxine-sensitive patients, the platelet survival returned to normal on treatment and in the non-responders it did not. In baboons experimental homocystinaemia did not alter platelet function but did reduce survival. Direct examination of the vessels of these animals showed patchy areas of endothelial loss. Circulating endothelial cells were sometimes found in blood samples and arterial thrombosis occurred. It was concluded that sustained endothelial injury due to homocystinaemia with the formation of platelet thrombi accounted for the reduced survival time and the subsequent atherosclerotic lesions. This study also suggested that thrombosis would be prevented by giving pyridoxine in those patients biochemically responsive to it; in non-responders it was suggested that dipyridamole (100 mg) and acetylsalicylic acid (up to 1 g) daily would provide protection against thrombus formation. However, Uhlemann et al., (1976) subsequently found that the platelet survival in 6 patients with homocystinuria (only 1 of whom was pyridoxine-responsive) was normal. Further, the platelets were apparently normal on electron microscopy.

The exact basis of the B_6 dependency in some forms of homocystinuria is biochemically interesting, and one may ask whether the administration of B_6 increases the synthesis of cystathionine synthase or reduces the rate of its breakdown.

CONGENITAL CONTRACTURAL ARACHNODACTYLY

The recorded features in this condition (Beals and Hecht, 1971; Lowry and Guichon, 1972) are contractures, long, thin extremities with arachnodactyly, scoliosis and abnormalities of the external ears. The condition is inherited as an autosomal dominant. Lens dislocation and vascular abnormalities are not described. Importantly, the contractures and disability are worse early in child-hood and tend to improve with increasing age. In a personally-studied family a mother and her two sons are affected. Both sons at the age of 5 and 8 years have fixed flexion of their knees and hips with contractures, and tend to walk on their toes, but the condition is not progressive. The mother walked in the same way as a child but at the age of 29 walks normally. She still has contractures of the fingers.

IDIOPATHIC SCOLIOSIS

Scoliosis commonly occurs during the period of rapid growth, particularly in females and is regarded as 'idiopathic'. This is the largest group of patients with scoliosis, and the known causes of scoliosis contribute only a fraction of the total. However there are several reasons to think that the common scoliosis of adolescents is not as 'idiopathic' as it seems.

Thus scoliosis so often occurs in the inherited disorders of connective tissue that it might be considered as a clinical marker of an inherited abnormality of collagen, and it is reasonable to postulate a similar abnormality in the 'idiopathic' group. Some evidence in support of this comes from the work of Zorab *et al.,* (1971) who showed increased hydroxyproline excretion in scoliotics compared with age-matched controls, and from Francis, Sanderson and Smith (1976) who demonstrated that the stability of the polymeric collagen from the skin of idiopathic scoliotics was abnormally low for age. They suggested that one of the factors contributing to scoliosis in adolescent girls was the increased instability of collagen at a time of rapid growth. That there is a generalized abnormality of growth in scoliotics was suggested by the data of Willner (1974) and supported by the later observations of Clark (1977), who was not however able to demonstrate any difference between the hydroxyproline excretion in 10 adolescent scoliotic girls and the hydroxyproline percentiles she had established for the normal population. Spencer (1977) found that the somatomedin level in scoliotic children was normal. So far as inheritance is concerned Wynne-Davies (1973) found a high incidence of affected relatives. Available results are in favour of a dominant inheritance with low penetrance, or multifactorial inheritance, but different surveys have given inconsistent results. These are further discussed in a recent symposium (Zorab, 1977).

THE EHLERS–DANLOS SYNDROME

General features and biochemistry

The different diseases which comprise this syndrome are all rare. However the severe variation may produce some of the most difficult problems found in medicine, and in some of them inherited defects in collagen have been identified. The general features are well described by McKusick (1972, 1976) and by Pinnell (1978); Tables 8.I and 8.III summarize the information on these disorders, and *Figure 8.1* shows which metabolic defects have been identified.

The defects in the E–D syndrome lead to physical abnormalities in many systems; these are particularly marked in the skin, musculoskeletal system, eye, vascular and gastrointestinal systems, but the pattern of involvement differs according to the particular type of disorder.

The skin provides the most striking changes. It may feel abnormal ('velvety'); it is hyperextensible, but returns to its previous shape after stretching, and it is fragile. Thus minor injury produces gaping wounds, often over exposed areas such as the knees. The skin in such places becomes thin, pigmented and parchment-like. Bruising is common, as is bleeding after minor surgical procedures. The joints are characteristically hyperextensible. All sorts of bizarre movements are described, and repeated joint dislocations occur. Kyphoscoliosis can develop. The eyes may show many abnormalities; blue sclerae, keratoconus, retinal detachment, angioid streaks and microcorneae are all described. Finally in some forms of the E–D syndrome extensive bruising with prolonged bleeding, or uncontrollable haemorrhage, or spontaneous rupture of the bowel can occur.

Different types of E–D syndrome

With the increasing recognition of the heterogeneity of this syndrome it is clear that certain clinical features are associated with particular biochemical defects. This is particularly so in Type IV, Type VI and Type VII E–D syndrome.

Type IV E–D syndrome

In this condition, which may itself be heterogeneous, an inherited inability to produce Type III collagen has been identified (Pope *et al.*, 1975). The clinical features which this produces are dramatic, and attest to the importance of Type III collagen normally present in skin and blood vessels (but not in bone) (p. 11). The skin is very pale and thin, and a prominent venous network is seen through it. There is severe bruising and particularly rupture of the bowel and large vessels.

Type VI E–D syndrome

This was the first heritable disorder of connective tissue in which a defect of collagen at a basic biochemical level was identified. It was first labelled as hydroxylysine-deficient collagen disease (Pinnell *et al.*, 1972) and later grouped

TABLE 8.III

The clinical features of the various types of Ehlers–Danlos syndrome. (*See* McKusick, 1976)

Type	Form	Main features	Mode of inheritance	Metabolic defect (proved or postulated)
I	Classic, severe form	Skin – stretchable, bruisable, fragile. Hyperextensible joints	Autosomal dominant	Unknown
II	Classic, mild form	Mild hypermobility and skin changes	Autosomal dominant	Unknown
III	Benign hypermobility syndrome	Marked mobility of joints		
IV	Echymotic, arterial or Sack-Barabas type	Thin skin, excessive bleeding, vessel rupture and gut rupture	Recessive	Does not form Type III collagen
V	X-linked form	Some have floppy mitral valves	X-linked recessive	Unknown*
VI	Ocular-scoliotic form	Fragility of eye globes Severe scoliosis Vascular abnormalities	Recessive	Lysyl hydroxylase deficiency
VII	Procollagen peptidase deficiency	Severe loose-jointedness	Recessive	Procollagen peptidase deficiency

* There is some evidence that this disorder is due to lysyl oxidase deficiency

with the E–D syndrome. The striking clinical features (Sussman *et al.,* 1974) are severe scoliosis, hyperextensibility with 'floppiness', and in particular abnormalities of the eye. These latter include small corneae with blue sclerae and fragility of the globe of the eye which easily ruptures. McKusick (1972) calls E–D VI the 'ocular-scoliotic' form.

Type VII E–D syndrome

In this condition, the main abnormality is severe loose-jointedness and multiple joint dislocation. Patients studied have been born with bilateral dislocation of the hips, were very floppy, bruised easily and had short stature. Lichtenstein *et al.,* (1973) showed an increased amount of procollagen in the skin and tendon compared with controls, together with a reduced activity of procollagen peptidase in cultured fibroblasts from affected patients.

ALKAPTONURIA

This is one of the classic inborn errors of metabolism, and since it appears to have an effect on connective tissues it is conveniently considered in this chapter. The metabolic block is produced by a deficiency of the enzyme homogentisic acid oxidase (*Figure 8.7*), normally responsible for the further metabolism of homogentisic acid formed in the breakdown of tyrosine. The clinical effects are produced by the accumulation of homogentisic acid (and its polymers) proximal to the block. The incidence of the disorder varies widely; it is inherited as a recessive and thus is seen more often in inbred communities with a high incidence of consanguinity. Alkaptonuria has significant effects on the skeleton. Its main characteristics are dark urine, pigmentation of connective tissues and a progressive and characteristic arthropathy. Other features are described by McKusick (1972).

Clinical features

Although alkaptonuria is rare, the diagnosis is easily made provided that it is thought of, particularly in a patient with unexplained arthritis. The term alkaptonuria is derived from early observations on the chemical reactions of the urine and ochronosis from the colour of the affected tissues, (*see* McKusick, 1972), but they both refer to the same disorder. The main abnormalities are found in the urine, the connective tissues and the joints.

Urine

The urine becomes dark on standing, and this darkening occurs more rapidly when the urine is alkaline. When passed it may be a normal colour. It will give a positive reaction when tested for reducing substances, but not with glucose oxidase. This may lead either to the erroneous diagnosis of diabetes mellitus

220

ALKAPTONURIA

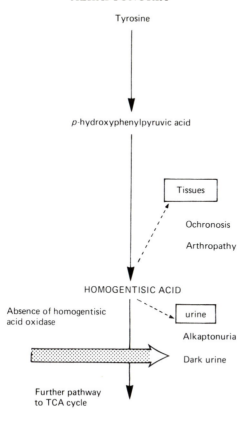

Tyrosine

↓

p-hydroxyphenylpyruvic acid

Tissues

Ochronosis

Arthropathy

HOMOGENTISIC ACID

Absence of homogentisic
acid oxidase

urine

Alkaptonuria

Dark urine

Further pathway
to TCA cycle

*Figure 8.7. The metabolic block in alkaptonuria
and its effects. The block is beyond homogentisic
acid*

from the use of Benedict's test or Clinitest tablets; or to failure to detect reducing substance if glucose oxidase (Clinistix) is used. Where the clinical manifestations suggest alkaptonuria, the darkening of the urine on the addition of alkali will rapidly confirm this.

Connective tissues

The eyes, ears and nasal cartilages become pigmented, with a blue-black appearance, as do joints and articular cartilage (seen at operation). Pigmentation in the sclerae can be well seen where the rectus muscles are inserted, and a similar pigmentation is described at the periphery of the cornea on the medial and lateral aspects.

The pinnae of the ears become stiff as well as pigmented, and the end of the nasal cartilage darkens. Pigmentation is also found in various secretions, particularly cerumen and sweat.

Arthropathy

Involvement of the joints and spine provide the main troublesome features of this disorder. In the spine there is early disc degeneration, with narrowing of disc spaces and calcification of many of the discs (*Figure 8.8*). The knees, hips

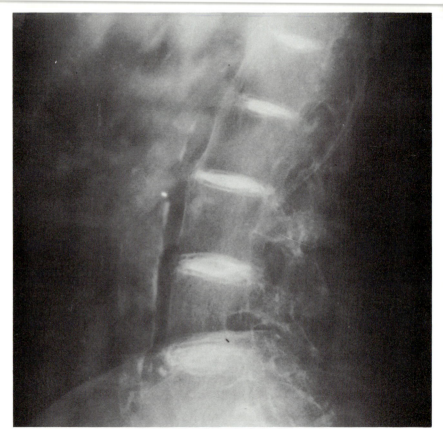

Figure 8.8 The appearance of the spine in alkaptonuria. The intervertebral spaces are narrowed and the intervertebral discs are calcified. (X-ray kindly supplied by Mr. A. G. Apley)

and shoulders show a premature degenerative arthritis. Thus the main presenting symptoms are pain in the back, and pain and limited movement in the larger joints.

Diagnosis

The diagnosis of alkaptonuria is often missed especially in a patient who does not constantly excrete dark urine. Thus it may be made first at an operation for arthritis, when tissues such as cartilage, tendons and ligaments are found to be darkly stained.

DISCUSSION

Treatment

Since it is considered that the deleterious effects of the disorder are in some way related to the accumulation of homogentisic acid and its derivatives, the logical medical treatment would be to reduce its precursors in the diet such as tyrosine, or less specifically, protein. There is little evidence that such measures are effective.

The effect of homogentisic acid on tissues

The relationship between homogentisic acid and the clinical features is not clear. Polymerized oxidized homogentisic acid is thought to increase the cross-linking of collagen by a process similar to the tanning of leather but the exact biochemical mechanisms are not known.

MENKES KINKY HAIR SYNDROME

This very rare condition is mentioned here since at least some of its manifestations are due to the effects of copper deficiency on collagen and elastin. The disorder is inherited as an X-linked recessive and there appears to be a block in the intestinal absorption of copper (Danks *et al.*, 1972). The intravenous infusion of copper may produce an increase towards normal of the depressed caeruloplasmin levels whereas oral copper does not. The clinical features are similar to some of those occurring in copper deficiency in cattle. Affected infants have abnormal 'kinky' hair, coarse features, impaired growth and central nervous system degeneration leading to early death. The hair is poorly pigmented and tangled or may be completely lost; microscopy of single hairs shows them to be twisted. Radiology shows multiple wormian bones and cupping of the ends of the long bones. The arteries are tortuous and irregular, and microscopy shows widespread fragmentation and reduplication of the elastic laminae.

DISCUSSION

The disease described in this chapter probably form the tip of the iceberg of inherited abnormalities of collagen (and elastin). In some of them (such as E–D IV) it is relatively easy to understand the clinical features in terms of the identified biochemical defect, but in others it is difficult to see why such differing tissues should be so variably affected, as for example those in the ocular-scoliotic form, E–D VI. In those disorders in which the primary defect appears to be in collagen synthesis, continuing biochemical advance is necessary to elucidate the mechanism responsible for the normal control of different genetic collagens. In other disorders such as homocystinuria and alkaptonuria (p. 220) the change in collagen is secondary to a primary biochemical defect, and the connection between them (excess of SH groups, deficiency of copper) is assumed on the basis of the known normal metabolic pathways of collagen. However, the effects of homocystine (and its monosulphide precursor) on the skeleton and other tissues is not fully understood.

In comparing the E–D syndromes with osteogenesis imperfecta (Chapter 7) it is interesting to note the very different effects of a complete deficiency of Type III collagen (E–D IV) and a relative lack of Type I (mild OI). It is unlikely that a disorder associated with complete inability to make any Type I collagen will be identified since it would probably be incompatible with life.

There are other inherited disorders of connective tissue which do not appreciably affect the skeleton and in which the biochemical cause is unknown; these are described in detail by McKusick (1972) and will not be dealt with further here. One of these is pseudoxanthoma elasticum.

SUMMARY

The inherited disorders of connective tissue such as Marfan's syndrome, E–D syndromes and disorders with similar clinical pictures are emerging from the status of curiosities to important diseases consequent upon identifiable biochemical defects. With the elucidation of their biochemical basis correct classification is becoming possible and particularly with the E–D syndromes this provides help in management and prognosis. It is important that every attempt should be made to establish the correct diagnosis in view of this increased knowledge. It is also important that a new look should be taken at 'idiopathic' scoliosis to identify those patients who may have unrecognized inherited biochemical abnormality.

REFERENCES

Beals, R. K. and Hecht, F. (1971). Congenital contractural arachnodactyly. *J. Bone Jt Surg.* **53A,** 987–993

Brenton, D. P. (1977). Skeletal abnormalities in homocystinuria. *Postgrad. Med. J.* **53,** 488–494

Brenton, D. P., Cusworth, D. C., Dent, C. E. and Jones, E. E. (1966). Homocystinuria. Clinical and dietary studies. *Q. Jl Med.* **35,** 325–346

Brenton, D. P., Cusworth, D. C., Biddle, S. A., Garrod, P. J. and Lasley, L. (1977). Pregnancy and Homocystinuria. *Ann. Clin. Biochem.* **14,** 161–162

Brenton, D. P., Dow, C. J., James, J. I. P., Hay, R. L. and Wynne-Davies, R. (1972). Homocystinuria and Marfan's syndrome. A comparison. *J. Bone Jt Surg.* **54B,** 277–298

Brill, P. W., Mitty, H. A. and Gaull, G. E. (1974). Homocystinuria due to cystathionine synthase deficiency. clinical–roentgenologic correlations. *Am. J. Roentg.* **121,** 45–54

Carson, N. A. J. and Neill, D. W. (1962). Metabolic abnormalities detected in a survey of mentally backward individuals in Northern Ireland. *Archs Dis. Childh.* **37,** 505–513

Carey, M. C., Donovan, D. E., Fitzgerald, O. and McAuley, F. D. (1968). Homocystinuria. 1. A clinical and pathological study of nine subjects in six families. *Am. J. Med.* **45,** 7–25

Clark, S. (1977). Longitudinal growth studies in normal and scoliotic children. In *Scoliosis. Proceedings of a 5th Symposium.* London. 1976. Ed. P. A. Zorab. pp. 165–180. London and New York: Academic Press

Cusworth, D. C. and Dent, C. E. (1969). Homocystinuria. *Br. med. Bull.* **25,** 42–47

Danks, D. M., Stevens, B. J., Campbell, P. E., Gillespie, J. M., Walker-Smith, J., Bloomfield, J. and Turner, B. (1972). Menkes kinky hair syndrome. *Lancet* **1,** 1100–1102

Dehnel, J. M. and Francis, M. J. O. (1972). Somatomedin (sulphation factor)–like activity of homocystine. *Clin. Sci.* **43,** 903–906

Field, C. M. B., Carson, N. A. J., Cusworth, D. C., Dent, C. E. and Neill, D. W. (1962). Homocystinuria. A new disorder of metabolism. (Abstract). In *Proceedings of the 10th International Congress of Pediatrics,* Lisbon. p. 274

REFERENCES

Fleisher, L. D., Longhi, R. C., Tallan, H. H., Beratis, N. G., Hirschhorn, K. and Gaull, G. E. (1974). Homocystinuria: investigations of cystathionine synthase in cultured fetal cells and the prenatal determination of genetic status. *J. Pediat.* **85**, 677–680

Francis, M. J. O., Sanderson, M. C. and Smith, R. (1976). Skin collagen in idiopathic adolescent scoliosis and Marfan's syndrome. *Clin. Sci. molec. Med.* **51**, 467–474

Francis, M. J. O., Smith, R. and MacMillan, D. C. (1973). Polymeric collagen of skin in normal subjects and in patients with inherited connective-tissue disorders. *Clin. Sci.* **44**, 429–438

Gaull, G., Sturman, J. A. and Schaffner, F. (1974). Homocystinuria due to cystathionine synthase deficiency: enzymatic and ultra-structural studies. *J. Pediat.* **84**, 381–390

Griffiths, R., Tudball, N. and Thomas, J. (1976). The effect of induced elevated plasma levels of homocystine and methionine in rats on collagen and elastin structures. *Connect. Tissue Res.* **4**, 101–106

Harker, L. A., Slichter, S. J., Scott, C. R. and Ross, R. (1974). Homocystinemia. Vascular injury and arterial thrombosis. *New Engl. J. Med.* **291**, 537–543

Kang, A. H. and Trelstad, R. L. (1973). A collagen defect in homocystinuria. *J. clin. Invest.* **52**, 2571–2578

Lichtenstein, J. R., Martin, G. R., Kohn, L. D., Byers, P. H. and McKusick, V. A. (1973). Defect in conversion of procollagen in a form of Ehlers–Danlos syndrome. *Science, N.Y.* **182**, 298–299

Lowry, R. B. and Guichon, V. C. (1972). Congenital contractural arachnodactyly; a syndrome simulating Marfan's syndrome. *J. Canad. Med. Ass.* **107**, 531–533

McKusick, V. A. (1972). *Heritable Disorders of Connective Tissue.* 4th edn. St. Louis: C. V. Mosby

McKusick, V. A. (1976). Heritable disorders of connective tissue. New clinical and biochemical aspects. In *Twelfth Symposium on Advanced Medicine.* Ed. D. K. Peters. p. 170–191. London: Pitman Medical

Mudd, S. H. and Levy, H. L. (1978). Disorders of transulfuration. In *The Metabolic Basis of Inherited Disease.* Eds J. B. Stanbury, J. B. Wyngaarden and D. S. Fredrickson. 4th edn. pp. 458–503. New York: McGraw-Hill

Murdoch, J. L., Walker, B. A., Halpern, B. L., Kuzma, J. W. and McKusick, V. A. (1972). Life expectancy and causes of death in the Marfan syndrome. *New Engl. J. Med.* **286**, 804–808

Pinnell, S. R. (1978). Disorders of collagen. In *The Metabolic Basis of Inherited Disease.* Eds. J. B. Stanbury, J. B. Wyngaarden and D. S. Fredrickson, 4th edn. pp. 1365–1394. New York:McGraw-Hill

Pinnell, S. R., Krane, S. M., Kenzora, J. E. and Glimcher, M. J. (1972). A heritable disorder of connective tissue. Hydroxylysine-deficient collagen disease. *New Engl. J. Med.* **286**, 1013–1020

Pope, F. M., Martin, G. R., Lichtenstein, J. R., Penttinen, R., Gerson, B., Rowe, D. W. and McKusick, V. A. (1975). Patients with Ehlers–Danlos syndrome. Type IV lack Type III collagen. *Proc. natn. Acad. Sci. USA* **72**, 1314–1316

Priest, R. E., Moinuddin, J. F. and Priest, J. H. (1973). Collagen of Marfan syndrome is abnormally soluble. *Nature, Lond.* **245**, 264–266

Sardharwalla, I. B., Fowler, B., Robins, A. J. and Komrower, G. M. (1974). Detection of heterozygotes for homocystinuria. *Archs Dis. Childh.* **49**, 553–559

Schimke, R. N., McKusick, V. A., Huang, T. and Pollack, A. D. (1965). Homocystinuria. Study of 20 families with 38 affected members. *J. Am. med. Ass.* **193**, 711–719

Sjoerdsma, A., Davidson, J. D., Udenfriend, S. and Mitoma, C. (1958). Increased excretion of hydroxyproline in Marfan's syndrome. *Lancet* **2**, 994

Spencer, G. S. G. (1977). Plasma and muscle somatomedin in scoliotic children. In *Scoliosis. Proceedings of a 5th Symposium.* London, 1976. Ed. P. A. Zorab. pp. 181–191. London and New York: Academic Press

Sussman, M., Lichtenstein, J. R., Nigra, T. P., Martin, G. R. and McKusick, V. A. (1974). Hydroxylysine-deficient skin collagen in a patient with a form of Ehlers–Danlos syndrome. *J. Bone Jt Surg.* **56A**, 1228–1234

Uhlemann, E. R., TenPas, J. H., Lucky, A. W., Schulman, J. D., Mudd, S. H. and Shulman, N. R. (1976). Platelet survival and morphology in homocystinuria due to cystathionine synthase deficiency. *New Engl. J. Med.* **295**, 1283–1286

Wilcken, B. and Turner, B. (1973). Homocystinuria. Reduced folate levels during pyridoxine treatment. *Archs Dis. Childh.* **48**, 58–62

225

REFERENCES

Willner, S. (1974). A study of growth in girls with adolescent idiopathic structural scoliosis. *Clin. Orthop. Related Res.* **101,** 129–135

Winter, R. B. (1977). The surgical treatment of scoliosis in Marfan's syndrome In *Scoliosis. Proceedings of a 5th Symposium.* London, 1976. Ed. P. A. Zorab. pp. 283–299. London and New York: Academic Press

Wynne-Davies, R. (1973). *Heritable Disorders in Orthopaedic Practice.* Oxford: Blackwell Scientific

Zorab, P. A. (1977). *Scoliosis. Proceedings of a 5th Symposium.* London, 1976. London and New York: Academic Press

Zorab, P. A., Clark, S., Cotrel, Y. and Harrison, A. (1971). Bone collagen turnover in idiopathic scoliosis estimated from total hydroxyproline excretion. *Archs Dis. Childh.* **46,** 828–832

226

9

Mucopolysaccharidoses

INTRODUCTION

There are a number of disorders affecting the skeleton which arise from failure of the normal breakdown of complex carbohydrates (Spranger, 1977). These carbohydrates accumulate in the tissues and appear in excess in the urine. These disorders are rare but demonstrate well the relation between inherited metabolic disease and the skeleton and require brief consideration in a book of this sort. They may be divided into two main groups, according to the nature of the substances stored, namely the mucopolysaccharidoses and the mucolipidoses. In the latter group lipids accumulate in addition to mucopolysaccharides and oligosaccharides appear in excess in the urine, providing the alternative term, the oligosaccharidoses. Spranger (1977) provides a useful short review of the mucolipidoses, and they will not be considered further here.

The mucopolysaccharidoses were first described as separate clinical entities (with names such as gargoylism) and only later recognized as disorders of mucopolysaccharide (glycosaminoglycan) storage. Initially they were identified by a combination of clinical features and urinary abnormalities but now the enzyme defects which cause most of these conditions have been recognized, and early attempts at enzyme replacement have been made (Di Ferrante *et al.*, 1971; Crocker, 1972). The development of knowledge about these disorders is well dealt with by McKusick (1972, 1976). Further reading should certainly include the classic work of Neufeld and her colleagues (Fratantoni, Hall and Neufeld, 1968; Weismann and Neufeld, 1970), who demonstrated clearly how the defects of fibroblasts from one type of mucopolysaccharidosis (MPS) could be mutually corrected by those from another type of mucopolysaccharidosis, and the review by McKusick, Neufeld and Kelly (1978).

Most people find that the mucopolysaccharidoses are more difficult to understand than those inherited disorders which involve collagen. This is because the chemistry of the mucopolysaccharides (glycosaminoglycans) is complex, because it is not understood how defects in their metabolism can be related to their clinical features, and because these clinical features overlap with a large

TABLE 9.1

The main mucopolysaccharide storage diseases. (Modified from Wynne-Davies and Fairbank, 1976. For further details *see* McKusick, 1976)

Designation		Inheritance	Excessive urine MPS	Enzyme-deficient	Main feature
MPS I H	Hurler syndrome	Autosomal recessive	Dermatan and heparan sulphate	α-L-iduronidase	Severe mental and skeletal changes. Death usually by 10 years
MPS II	Hunter syndrome	X-linked recessive	Dermatan and heparan sulphate	Iduronate sulphatase	Less severe than Hurler. All male
MPS III	Sanfilippo syndrome	Autosomal recessive	Heparan sulphate	Heparan N-sulphatase or N-acetyl α-D-glucosaminidase	Mental retardation. Little skeletal change
MPS IV	Morquio syndrome	Autosomal recessive	Keratan sulphate	?	Severe dwarfing, joints lax, intelligence normal
MPS V	Vacant				
MPS VI	Maroteaux-Lamy syndrome	Autosomal recessive	Dermatan sulphate	Aryl sulphatase B	Severe skeletal change. Normal intelligence

number of other diseases such as the spondylo-epiphyseal dysplasias and mucolipidoses.

Biochemically the mucopolysaccharidoses differ in an important respect from the inherited disorders of collagen which we have previously considered in that they are disorders of breakdown rather than of synthesis.

In this short chapter it would be inappropriate to deal with all the known mucopolysaccharidoses. The aims are to summarize the biochemical background, to relate it to the commoner forms of mucopolysaccharidosis which affect the skeleton and to consider their diagnosis and possible treatment. Table 9.I outlines current knowledge of the mucopolysaccharide storage disorders.

BIOCHEMISTRY

The mucopolysaccharides (Chapter 1) are very large molecules with a protein core to which are attached polysaccharide chains. Their precise chemistry is complex and understanding of the disorders which arise from their defective degradation is much helped by the straightforward explanation of McKusick (1976). He points out that all the mucopolysaccharidoses are lysosomal storage disorders in which the 'undigested or incompletely digested mucopolysaccharides' accumulate in the lysosomes, which are the 'garbage disposal' apparatus of the cell. The mucopolysaccharides progressively accumulate in many tissues, and are excreted in excess in the urine. The enzyme defects which cause specific mucopolysaccharidoses are shown in Table 9.I and *Figure 9.1* (from McKusick, 1976). These defects are concerned with the polysaccharide chains and not with

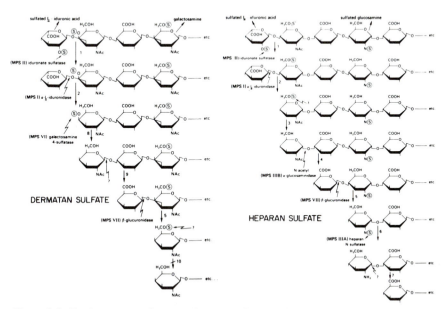

Figure 9.1. To demonstrate the normal steps in the degradation of dermatan and heparan sulphate, with the position of the enzyme deficiencies in the mucopolysaccharidoses. The chemistry is complex, but the result is always the accumulation of precursors proximal to the block. (See McKusick, 1976, for the larger original illustration)

the protein core. Again McKusick (1976) points out that enzyme degradation works like a zipper, 'if the enzyme that catalyses a particular step is missing, the zipper sticks and moves no further'.

When fibroblasts from one type of mucopolysaccharidosis are cultured with those from another type, or with normal fibroblasts, there may be correction of the mucopolysaccharide accumulation in the defective fibroblast. This correction is considered to be due to replacement of the missing enzyme, and the 'corrective factors' of Neufeld and Fratantoni (1970) contain the enzymes whose deficiency causes the specific type of mucopolysaccharidosis.

PATHOLOGY

The accumulation of the mucopolysaccharide before the metabolic block occurs within the lysosomes of all sorts of cells and thus very many tissues are affected and the disorders are on the whole progressive. Initially the stored material was thought to consist of lipid substances, but it is now clear that they are sulphated mucopolysaccharides. It is not clear why the accumulation of different mucopolysaccharides causes such different clinical and pathological pictures. We may take as an example the changes described in the Hurler syndrome (MPS I H). In this syndrome similar abnormalities have been found in many different cells which include those of fibroblastic origin, those of bone — the chondrocytes and osteocytes, the Kupffer and parenchymal cells of the liver, and the nerve cells of the central nervous system. Such cells are distended by large amounts of mucopolysaccharides and have been called 'gargoyle' cells. Electron microscopy of such cells and of cultured skin fibroblasts shows the storage of this material in enlarged lysosomes (*see* Figure 53–1, McKusick, Neufeld and Kelly, 1978).

CLINICAL FEATURES

The description will be limited to the Hurler, Hunter and Morquio syndrome (MPS IH, II and IV, respectively). In the Hurler syndrome the non-skeletal manifestations are severe and progressive, with mental deterioration and early death; by contrast in patients with the Morquio syndrome skeletal features predominate. Spranger (1977) emphasizes that with few exceptions the bone changes do not permit precise diagnosis of the type of dysplasia present.

The Hurler syndrome (MPS IH)

Of the many forms of MPS, this is the most severe and causes death at an early age. All patients with the disorder look the same, and have the features to which the out-dated term gargoylism has been applied. The enzyme defect is α-L-iduronidase (*Figure 9.1*) which normally splits the L-iduronate radical from the sulphated carbohydrates dermatan and heparan sulphate. When this enzyme is deficient there is an intracellular and intralysosomal accumulation of short-chain mucopolysaccharides with L-iduronate at the non-reducing end. Heparan and dermatan sulphate accumulate in the cells and when released appear in excess in the urine. Two other disorders MPS IS (Scheie syndrome) and MPS IH/S

(Hurler/Scheie syndrome) have the same enzyme defect but different clinical features. These may arise from different mutant forms of the same enzyme.

MPS IH is inherited as an autosomal recessive. Affected infants appear to develop normally for a few months after birth, but soon deteriorate mentally and physically. Death occurs in late childhood and is often due to pneumonia or to coronary artery disease associated with deposits of the mucopolysaccharides. The physical features include dwarfism, a typical facial appearance together with rhinitis and snuffling, a short neck with lumbar gibbus and chest deformity, and an enlarged abdomen.

The head and face are remarkable. The vault of the skull may be an unusual shape with scaphocephaly or acrocephaly, or variations on this, the lips and tongue are large, the mouth open, the bridge of the nose flattened and the features coarse and ugly. The gums may be hypertrophied over the enlarged alveolar ridges. Noisy breathing is very common as is variable deafness. The eyes are often very prominent. Corneal clouding is present. The hands are broad, the fingers stubby and the joints stiff. This stiffness extends to other joints. The abdomen is protruberant, with hepatosplenomegaly, and herniae may occur.

Radiology (see Wynne-Davies and Fairbank, 1976) demonstrates the abnormal shape of the skull (which may be thickened) the enlarged shoe-shaped sella turcica, the beaking of the vertebrae, the thoracolumbar kyphosis and the abnormal shape of the long bones and the phalanges. Radiological signs are seen in the vertebral column within the first few months of life. The infantile biconvex shape may persist longer than normal, and the last thoracic and first lumbar vertebrae are unduly small. The anterosuperior part of the vertebrae does not develop properly so that the inferior part projects like a blunt hook. Other changes are described in the pelvis, thorax and long bones. Wynne-Davies and Fairbank (1976) provide some particularly useful diagrams. The acetabular roof is oblique, associated frequently with subluxation of the hips. The ribs widen anteriorly giving a sabre or paddle-shaped appearance. The long bones show a modelling defect sometimes with relative wideness of the diaphyses. The hands show short phalanges and the metacarpals have coarse trabeculation. The metacarpals, except for the first, are particularly unusual in shape, with pointed proximal ends; the phalanges are narrower distally and referred to as 'bullet-shaped'.

Some radiological features of a personal case are shown in *Figure 9.2*. The pedigree (*Figure 9.3*), is compatible with recessive inheritance. Her parents had particularly noted the deformity of the spine, the abnormal shape of the head, the unusual appearance and continual 'snuffling'. She was first seen at the age of 3½ years and over the next two years deteriorated strikingly in appearance and mentality. The CPC (cetylpyridinium chloride) screening test (Manley and Williams, 1969) on the urine demonstrated excessive mucopolysaccharides which were mainly composed of dermatan sulphate and heparan sulphate.

The Hunter syndrome (MPS II)

Classically this is a milder form of MPS than the Hurler syndrome. There are also severe and mild forms of the disorder itself, probably caused by allelic mutations on the X chromosome. It is inherited as an X-linked recessive, and would therefore not be transmitted from father to son (see McKusick, 1972), but reproduction

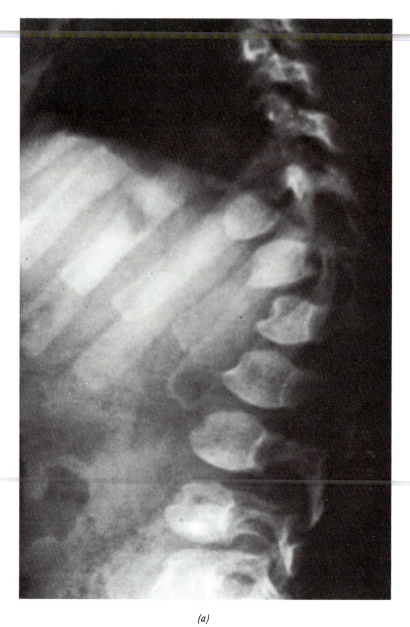

(a)

Figure 9.2. Some radiological features of a girl of 3 with Hurler's syndrome (MPS IH). (a) Shows the thoracolumbar gibbus, with hook-shaped vertebrae and persistence of the infantile shape of the vertebral bodies. The ribs are 'sabre'-shaped

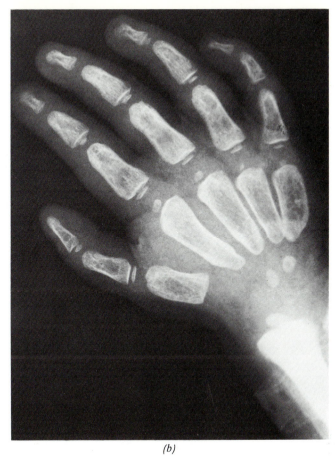

(b)

Figure 9.2 (b) Shows the proximal narrowing of metacarpals 2–5 and the 'bullet'-shaped phalanges

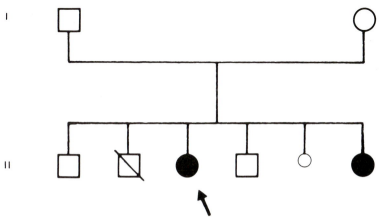

Figure 9.3. Pedigree of the patient in Figure 9.2. showing an affected sibling. Compatible with recessive inheritance. ● Affected females

233

is very rare. All patients are male. It is due to a deficiency of the enzyme iduronate sulphatase which is responsible for the removal of the sulphate radical, the first step in the degradation of the MPS (*Figure 9.1*), and there is an excess of dermatan and heparan sulphate in the urine. It differs from the Hurler syndrome in being less severe, both physically and mentally. A thoracolumbar gibbus does not occur and corneal clouding is not seen. Progressive deafness is said to be a feature, and also nodular thickening over the skin of the upper arms or thorax. *Figure 9.4* from a boy of 5 years with probable severe Hunter

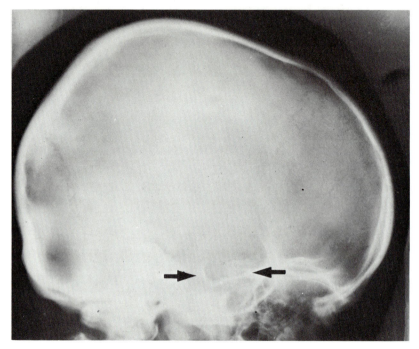

Figure 9.4. The radiological appearance of the skull in a patient with probable Hunter's syndrome. The x-ray is not well centred, but the pituitary fossa shows a characteristic 'shoe' shape

syndrome, shows that the changes in the pituitary may be the same as those of MPS I. At the age of 9 years he was on the 3rd percentile for height, mentally retarded with short stubby hands which were stiff and contracted, with no clinical evidence of corneal clouding and without any spinal deformity. Marked aortic incompetence was an unusual feature.

The Morquio syndrome (MPS IV)

The Morquio syndrome is probably best known for its orthopaedic manifestations. It is likely to be a heterogeneous disorder since only a proportion of patients excrete excessive amounts of keratan sulphate in the urine (hence an alternative term for the disorder is 'keratan sulphaturia'). Many different disorders have previously been included under this heading, including some of the spondylo-epiphyseal dysplasias, and it has probably been overdiagnosed. The difficulty in correct diagnosis will remain until the specific enzyme defects are recognized.

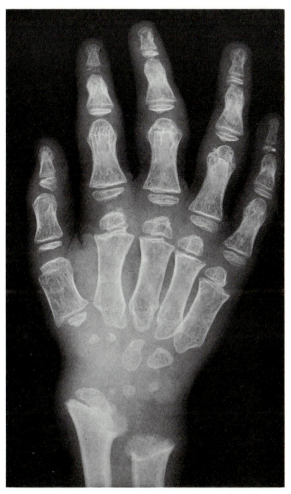

Figure 9.5. The radiographic appearance of the hand in a boy with the Morquio syndrome (MPS IV) whose growth chart is shown in Figure 9.6. In contrast to Figure 9.2. (b) the meta-carpals show constriction of the diaphyses

The severity of this disorder varies. Intelligence is normal and the major abnormal features are skeletal, which are progressive from birth. Within the first few years of life the child becomes deformed and dwarfed. There is a short neck, barrel chest with prominent sternum, flexed stance and often knock knees. The facial features are not so abnormal as in MPS IH but corneal clouding does occur. In contrast to MPS IH, the joints of patients with MPS IV can be excessively mobile and the skin itself is loose. Radiology during infancy shows an appearance in the thoracolumbar region similar to the Hurler syndrome, but later the main feature is platyspondyly, which contributes to the very short trunk, with a central anterior beak on the vertebrae. The carpal bones are abnormal, being small and irregular, and the metacarpals and phalanges have a central constriction (*Figure 9.5*). Importantly the odontoid is hypoplastic, failing to fuse normally with the body of C2 producing atlantoaxial instability; spinal cord compression may also result and is said to be present to a variable degree in most older patients with Morquio syndrome. The pelvis shows severe hypoplasia, with coxa valga and epiphyseal dysplasia. A boy of 10 years, whose growth chart is shown (*Figure 9.6*) and who is now of apparently normal intelligence,

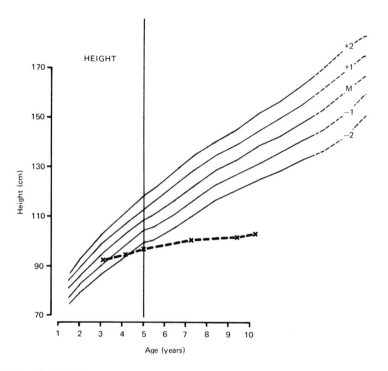

Figure 9.6. To demonstrate the progressive growth failure in Morquio's syndrome

demonstrates these features well. His chest deformity is progressive, he has severe knock knees with very hypermobile joints, his skin appears loose, thick and abnormal and he has corneal clouding. There is excessive urinary excretion of keratan sulphate.

The individual prognosis for a patient with MPS IV is variable. Death may occur early from cardiorespiratory failure and also from atlantoaxial dislocation. Progressive neurological disability with spastic paraplegia may occur. In patients who are having corrective orthopaedic operations for deformity it is important to remember that intubation during anaesthesia is potentially dangerous.

In addition to the possibility of medical treatment (p. 237) it has been suggested that elective cervical spinal fusion may improve prognosis.

DIAGNOSIS

The clinical and radiological features of a patient with a mucopolysaccharidosis or mucolipidosis are usually sufficient to suggest the diagnosis. Radiologically the bone changes come under the heading of 'dysostosis multiplex' which include the features described in this chapter under MPS IH.

It is important to exclude other conditions which superficially resemble MPS, particularly hypothyroidism (Chapter 2) in which the large tongue, coarse facial features, delayed skeletal growth and radiological abnormality may cause temporary confusion.

Providing that the main diagnosis of MPS (or mucolipidosis) is correct, identification of the specific disorder is helped by the flow-chart provided by Spranger, Langer and Wiedemann (1974). According to these authors, the radiological features of MPS IV, and mucolipidosis III are sufficiently specific to make the diagnosis. The remainder who have a common pattern of radiological change may be divided according to whether there is an excess of mucopolysaccharides in the urine (the mucopolysaccharidoses) or not (the mucolipidoses), and further subdivided according to severity and time of onset (infancy or childhood).

Biochemical confirmation of the diagnosis of a mucopolysaccharidosis is based on the presence of excess mucopolysaccharide in the urine and its precise identification; other approaches include the demonstration of various types of cellular inclusions and the correction of enzyme defects by the Neufeld type of experiment with mixed fibroblast cultures (*see* McKusick, Neufeld and Kelly, 1978). In practice, one of the simple screening tests for excess mucopolysaccharides in the urine is the first step.

Tests for the excess acid mucopolysaccharide excretion in the urine (Dorfman and Matalon, 1972) take advantage of the fact that they are large polyanionic macromolecules. Thus they may be precipitated by polycations (such as cetylpyridinium chloride (CPC) or cetyltrimethyl ammonium bromide, (CTAB)), or separated by chromatography on ion-exchange resins, or by various forms of electrophoresis. Methods of screening may employ the precipitating effect of acid mucopolysaccharides on albumin, or the turbidity produced by mixing mucopolysaccharides and CPC. The increased excretion of mucopolysaccharides is not limited to the mucopolysaccharidoses, and since it may be increased after injury, in infancy, and during normal growth, sensitive screening tests may give false positive results. Amongst the MPS disorders only patients with Hurler syndromes excrete enough mucopolysaccharide in the urine for it to be always detectable, and the definite identification of a specific type of MPS from the urinary findings can be complicated.

TREATMENT

The mucopolysaccharidoses provide difficult therapeutic problems. Before the underlying metabolic defects were recognized, treatment for symptoms and particularly skeletal deformity was all that was possible. Thus knock knee, as in the Morquio syndrome, might be corrected orthopaedically, and stabilization of the cervical spine considered. Now that the enzyme defects are recognized two problems arise, namely the possibility of antenatal diagnosis (Cantz and Gehler, 1976) and of enzyme replacement in affected individuals.

MUCOPOLYSACCHARIDOSES

The crudest form of enzyme replacement is provided by transfusion of fresh plasma or whole blood, but despite initial enthusiasm (Di Ferrante *et al.*, 1971) there is no convincing evidence that it is effective (Crocker, 1972). A more sophisticated approach is necessary, similar to that discussed by Gregoriadis (1976), in which enzymes are transported in liposomes to their site of action. The antenatal diagnosis of a mucopolysaccharidosis is now possible in specialized centres (McKusick, Neufeld and Kelly, 1978).

SUMMARY

The mucopolysaccharidoses are rare disorders due to failure of mucopolysaccharide degradation and consequent accumulation in tissues. Although many of the responsible enzyme defects have been identified, the ways in which they produce the differing clinical features are not understood. Practically it is important to obtain as accurate a diagnosis as possible, since prognosis and possible future treatment may depend on it.

REFERENCES

Cantz, M. and Gehler, J. (1976). The mucopolysaccharidoses: inborn errors of glycosaminoglycan metabolism. *Hum. Genetics* **32**, 233–255

Crocker, A. C. (1972). Plasma infusion therapy for Hurlers' syndrome. *Pediatrics* **50**, 683–685

Di Ferrante, N., Nichols, B. L., Donnelly, P. V., Giovanni, N., Hrgovcic, R. and Berglund, R. K. (1971). Induced degradation of glycosaminoglycans in Hurler's and Hunter's syndromes by plasma infusion. *Proc. natn Acad. Sci. U.S.A.* **68**, 303–307

Dorfman, A. and Matalon, R. (1972). The mucopolysaccharidoses. In *The Metabolic Basis of Inherited Disease.* Eds. J. B. Stanbury, J. B. Wyngaarden and D. S. Fredrickson. 3rd edn. New York: McGraw Hill

Fratantoni, J. C., Hall, C. W. and Neufeld, E. F. (1968). Hurler and Hunter syndromes: Mutual correction of the defect in cultured fibroblasts. *Science, N. Y.* **162**, 570–572

Gregoriadis, G. (1976). The carrier potential of liposomes in biology and medicine. *New Engl. J. Med.* **295**, 704–710, 765–770

Manley, G. and Williams, U. (1969). Urinary excretion of glycosaminoglycans in the various forms of gargoylism. *J. clin. Path.* **22**, 67–75

McKusick, V. A. (1972). *Heritable Disorders of Connective Tissue.* 4th edn. St. Louis: C. V. Mosby Co

McKusick, V. A. (1976). Heritable disorders of connective tissue. New clinical and biochemical aspects. In *Advanced Medicine* Vol. 12. Ed. D. K. Peters pp. 170–191 London: Pitman Medical

McKusick, V. A., Neufeld, E. F. and Kelly, T. E. (1978). The mucopolysaccharide storage diseases. In *Metabolic Basis of Inherited Disease.* Eds J. B. Stanbury, J. B. Wyngaarden and D. S. Fredrickson. 4th edn. New York: McGraw Hill

Neufeld, E. F. and Fratantoni, J. C. (1970). Inborn errors of mucopolysaccharide metabolism. Faulty degradative mechanisms are implicated in this group of human diseases. *Science N. Y.* **169**, 141–146

Spranger, J. W. (1977). Catabolic disorders of complex carbohydrates. *Postgrad. med. J.* **53**, 441–448

Spranger, J. W., Langer, L. O. and Wiedemann, H. R. (1974). Bone Dysplasias. An atlas of constitutional disorders of skeletal development. London: W. B. Saunders Co

Weismann, U. and Neufeld, E. F. (1970). Scheie and Hurler syndromes: apparent identity of the biochemical defect. *Science N. Y.* **169**, 72–74

Wynne-Davies, R. and Fairbank, T. J. (1976). Fairbanks Atlas of General Affections of the Skeleton. 2nd edn. London and Edinburgh: Churchill Livingstone

10

Ectopic Mineralization

INTRODUCTION

The deposition of mineral in the soft tissues, either as calcification or ossification, provides a number of challenging physiological and clinical problems. In some of them the cause appears obvious, but in others it is quite unknown. Since they involve bone mineral and may be associated with disorders of the skeleton, as well as the formation of bone outside the skeleton, they require brief consideration in this book. There are several classifications, according to supposed differences in aetiology and pathology, and one practical division is suggested in Table 10.I. Clinically it is important to distinguish those conditions (mainly acquired) in which the biochemistry is abnormal and there is some prospect of altering the course of the disorder, from those which appear to follow some form of tissue damage, and from those which are inherited. It is also important to distinguish whether or not mineralization in the soft tissue is associated with ectopic bone matrix (myositis ossificans).

These are disorders in which advance has been slow, despite some clues. In myositis ossificans the major problem is what causes the bone matrix to form in the soft tissues in the first place and how this can be prevented. Therapeutically the diphosphonates (Russell, 1975) have not fulfilled the promise suggested by the prevention of induced ectopic calcification in animals.

CALCIFICATION SECONDARY TO BIOCHEMICAL CHANGES

Where the circulating level of calcium and/or phosphate is consistently high (giving a high CaXP product) mineralization can occur in many soft tissues. The distribution of the mineral varies according to its cause, and includes the blood vessels, the skin, the cornea, the conjunctivae, the brain and the kidneys. Thus in primary hyperparathyroidism or in hyperparathyroidism associated with renal failure, calcification occurs in the blood vessels and periarticular soft tissues whereas in parathyroid insufficiency it is particularly seen in the subcutaneous tissues and the brain (Chapter 6).

TABLE 10.I
Some causes of ectopic mineralization

A. *Calcification secondary to biochemical abnormalities**

  (1) Hyperparathyroidism (high calcium: other causes include vitamin D overdose, milk alkali syndrome, sarcoidosis)

  (2) Renal failure (high phosphate, sometimes high calcium)

  (3) Idiopathic hyperphosphataemia (tumoral calcinosis; high phosphate)

  (4) Hypoparathyroidism and pseudohyperparathyroidism (high phosphate)

B. *Calcification secondary to tissue 'damage'*

  (1) After dermatomyositis (may be generalized: 'calcinosis universalis')

  (2) With scleroderma

  (3) After infection (such as tuberculosis)

  (4) Cause unknown (may be localized: 'calcinosis circumscripta')

C. *Ossification*

  (1) Acquired. After trauma, surgery or paraplegia

  (2) Inherited: Myositis ossificans progressiva

D. *Miscellaneous rare causes*

  (1) With pseudoxanthoma elasticum

  (2) Around tendons: 'calcific tendinitis'

* In the plasma

In one particular form of ectopic calcification masses of calcium accumulate around the large joints. This is sometimes referred to as 'tumoral calcinosis', a term which has been applied to ectopic calcification of this distribution occurring either without detectable biochemical abnormality, or with naturally-occurring or induced increases in circulating calcium or phosphate, or with inherited hyperphosphataemia. Walker *et al.,* (1977) give a good example of tumoral calcinosis resulting from treatment with lα-hydroxycholecalciferol. However it is probably preferable to restrict the term tumoral calcinosis to the rare recessively-inherited disorder characterized by an isolated increase in plasma phosphate (Lafferty, Reynolds and Pearson, 1965; Wilber and Slatopolsky, 1968; Baldursson *et al.,* 1969). In this condition the plasma phosphate may be increased to about twice normal, due to a considerable increase in renal-tubular phosphate re-absorption. Masses of calcium appear in the soft tissues in childhood and may require surgical removal. They tend to be partially encapsulated and occur in relation to the larger joints. *Figure 10.1* demonstrates the radiological appearance in a 26-year-old man, who first developed calcification within the muscles over the right hip at about the age of 6, and had a plasma phosphate of 8.4 mg

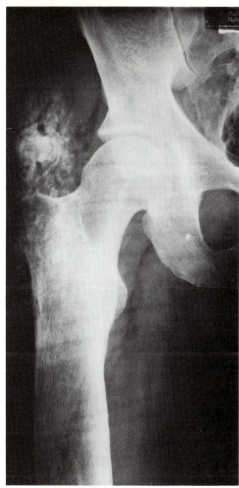

Figure 10.1. Ectopic calcification in the periarticular tissues in a man of 26 years with tumoral calcinosis, who had symptoms from the age of 6 years. Unexpectedly the shape of the femur and other bones was also abnormal. Oral aluminium hydroxide (30 ml four times a day) reduced his plasma phosphate from 8.4 to 5.6 mg per 100 ml

per 100 ml (2.7 mmol/l). There was a family history of affected members and probable consanguinity.

Treatment aims at reducing the plasma phosphate, either by decreasing phosphate intake with aluminium hydroxide (Mozaffarian, Lafferty and Pearson, 1972) or decreasing its tubular reabsorption with probenecid.

CALCIFICATION SECONDARY TO TISSUE DAMAGE

With the exception of insignificant local calcification such as that which may follow tuberculosis, infected pleural effusions or local bleeding, there appear to be two fairly distinct forms of more generalized calcification which follow either dermatomyositis or scleroderma. These can be considered as forms of calcinosis in association with acquired connective tissue (so-called 'collagen') diseases. They may also be considered examples of dystrophic calcification, but this concept adds nothing of practical use.

241

Dermatomyositis

The extensive soft tissue calcification in dermatomyositis (*Figure 10.2*) follows some years after the first onset of the disease, and often in late childhood. There may be a previous history of a generalized illness, with painful weak proximal muscles and inability to walk, and often a rash. The initial diagnosis is based on

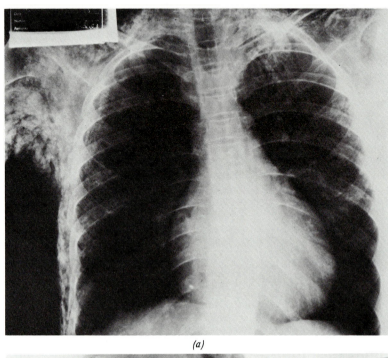

(a)

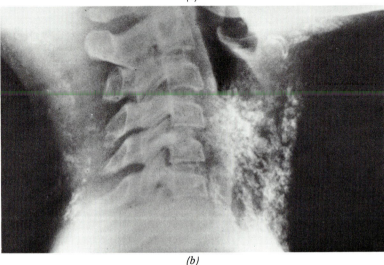

(b)

Figure 10.2. The appearance of ectopic calcification following dermatomyositis.
(a) and (b) as plaques or sheets of mineral under the skin ('calcinosis universalis')

242

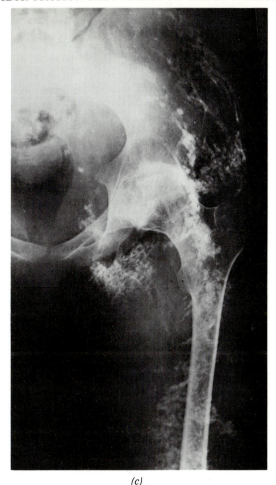

(c)

*Figure 10.2. (c) Additionally some calcification appears
to follow the connective tissue planes of muscle*

the clinical picture, the muscle biopsy findings and the increased phosphocreatinine kinase in the plasma. Typically, by the time the patient is seen, steroids and/or immunosuppressive drugs have been given with slow improvement. It is during this phase that calcinosis may first begin. It may form as sheets of calcium beneath the skin and can then be referred to as 'calcinosis universalis' but again this name, and 'calcinosis circumscripta' (for localized calcinosis) is purely descriptive and does not help towards the aetiology. The mineral may also appear to be aligned along the fibrous planes within muscles such as the quadriceps and glutei. Calcification of this sort after dermatomyositis does not necessarily break through the skin but can be easily felt beneath it. Occasionally it may accumulate, particularly around the joints and extremities, in the way seen in scleroderma.

Scleroderma

A more common form of ectopic calcification is that in association with scleroderma (*Figure 10.3*) and the calcinosis may occur with Raynaud's phenomena, scleroderma and telangiectases (CRST syndrome). The ectopic mineral accumulates subcutaneously around the phalanges and the peripheral joints and also over pressure areas, such as knees, elbows and buttocks. Considerable amounts may eventually be present and from time to time material with the

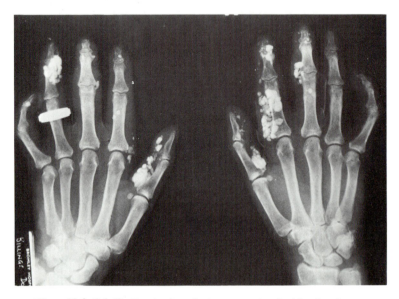

Figure 10.3. Calcification in the soft tissues associated with scleroderma

consistency of toothpaste discharges through the skin. X-ray diffraction can be used to demonstrate its composition which is that of hydroxyapatite. Extensive ectopic calcification can also occur without any evidence of previous or present disease. In the girl whose radiographs are shown in *Figure 10.4* discharge of the ectopic mineral through the skin was associated with marked fever.

Treatment

There is no proven beneficial treatment for these forms of calcinosis. Luckily those which occur in childhood in association with dermatomyositis may rapidly (and unpredictably) improve, sometimes with the onset of puberty. Attempts at medical treatment include the use of phosphonates, of EDTA (to increase calcium excretion), of aluminium hydroxide (to reduce phosphorus intake) and of probenecid (to reduce plasma phosphate concentration). Although it has been demonstrated that diphosphonates can prevent ectopic calcification induced in animals, the use of EHDP (Chapter 5) in ectopic mineralization in man has not been predictably beneficial, and where the calcification has disappeared it has been difficult to be certain that this would not have occurred naturally. Limited trials show no striking effect in the calcinosis of scleroderma. Nassim

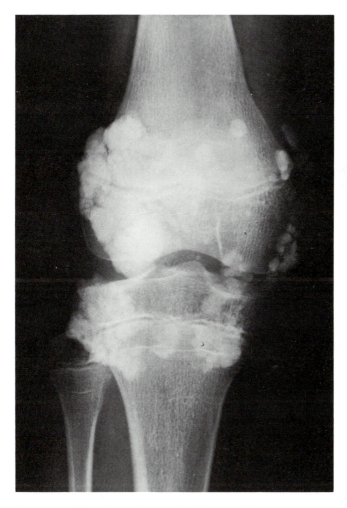

Figure 10.4. 'Idiopathic' calcinosis. In this girl of 10 years there was no history of previous illness

and Connolly (1970) reported the beneficial use of aluminium hydroxide in calcinosis universalis, and Dent and Stamp (1972) demonstrated a considerable improvement in calcinosis circumscripta in a female treated with oral probenecid, but others have found that this drug is not well tolerated and is without striking effect. Since the plasma phosphate is normal in these forms of calcinosis, consistent benefit would be unexpected.

245

OSSIFICATION

In ectopic ossification true bone forms outside the skeleton and often within the connective tissues of muscle. The cause is unknown and it is not always clear where the process first starts. However the first important change is the appearance of extraskeletal bone matrix, and this subsequently mineralizes. Attempts to prevent mineralization are important, but do not cope with the underlying disorder. The major disability results from the progressive fixation of joints.

Acquired ectopic ossification

Acquired ectopic ossification in association with tissue damage may follow accidental injury or elective surgery, including total hip replacement. Ossification is also a feature of neurological injury and particularly paraplegia. The cause of ectopic ossification after paraplegia is quite unknown, but it has been reported to occur in up to half such patients, and to affect predominantly the tissues around the hip joints. It is only in the minority that severe limitation of movement occurs.

In the treatment of these forms of ossification, the diphosphonate EHDP (Chapter 5) has again been tried. A randomized study of patients after hip replacement (Bijvoet *et al.*, 1974) suggested that if EHDP was given immediately after operation and before there was any detectable periarticular calcification, it would improve the mobility of the joint and reduce the expected incidence of mineralization. However, when EHDP was stopped ectopic mineral appeared in the same number of treated patients as of untreated controls. This finding implied that EHDP did not prevent but merely delayed the deposition of calcium in the ectopic bone matrix, but the advantage in mobility appeared to be maintained.

Similar results were obtained by Stover, Niemann and Miller (1976) who studied the effect of EHDP in preventing the recurrence of ectopic ossification around the hip after its surgical removal from four patients with spinal cord injuries. Two patients, in whom ossification had recurred after previous operations without EHDP, acted as their own controls; in two other patients, one was given EHDP and one was given placebo before, during and after bone resection. Without EHDP ectopic bone resection was followed by bony ankylosis, whilst EHDP modified or prevented it. Ectopic mineral reappeared after EHDP was stopped but the original postoperative gain in mobility continued. However as in MOP (p. 251) there were individual differences in response, and prolonged postoperative treatment with EHDP appeared to be necessary.

Inherited – myositis ossificans progressiva (MOP)

This very rare disorder – also known as fibrodysplasia ossificans progressiva (McKusick, 1972) – has for long been a clinical and biochemical mystery. In this condition the diagnosis is made by the tendency towards ossification in the muscles combined with characteristic skeletal abnormalities. The condition has been reviewed by Lutwak (1964), McKusick (1972) and Smith (1975).

246

Clinical features

The typical skeletal abnormality, present at birth, is a short big toe (*Figures 10.5* and *10.6b*). The ossification centres are abnormal and if the deformity is noted, hallux valgus may be diagnosed (and may be operated upon). Later the big toe appears to be monophalangic, but many variations exist (*Figure 10.5*). Characteristically the thumbs also are short but this is less easy to assess than the

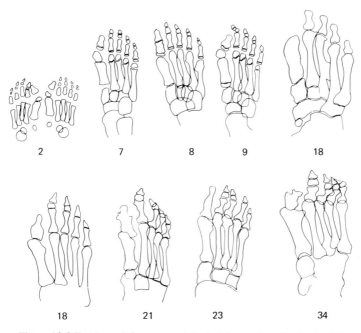

Figure 10.5 Tracings of the x-rays of the feet in a series of patients with MOP. The age of the patient in years is shown beneath each x-ray.
(From Smith, 1975)

short big toes. However the deformity of the hands may be considerable (Smith, Russell and Woods, 1976, Figures 10 and 11). Although the phalangeal abnormalities are usually easily recognizable, there are other less obvious skeletal changes such as abnormal shape of the long bones and exostoses. A striking feature which may falsely suggest Still's disease is fusion of the cervical vertebrae which appears to start early in life and cannot always be attributed to prior ossification within the muscles of the neck (*Figure 10.6*).

Myositis (so-called) may begin within the first year or so, and most often around the neck and upper back. Typically the affected muscles become painful, red and hard, and this may be attributed to bruising, infection or even to sarcoma. Doubt about the diagnosis may lead to biopsy, which will reveal much destruction of myofibrils associated with cellular infiltration. The area of muscle clinically involved may be large and associated with systemic upset. After a week or so the swelling and pain subside and improvement is often thought to have occurred; however there is soon radiological evidence of progressive mineralization which forms blocks or bars of bone, and fixes joints.

Intermittent attacks of myositis, with progressive ossification, continue throughout childhood although there may be relatively long periods of quiescence. Movements of the neck, shoulders and chest become very limited, but the most catastrophic disability is produced by ossification of the muscles around the hip, when walking becomes impossible.

Although the intermittent 'myositis' is characteristic, another form of presentation is progressive stiffness. In such patients ossification may be late and the diagnosis missed, especially if the significance of the monophalangic big toe is not recognized.

Routine biochemical measurements are normal, with the exception of the plasma alkaline phosphatase which may be increased during active myositis.

Cause

MOP is usually classified as a disorder of connective tissue, inherited as an autosomal dominant trait with variable penetrance. Mild skeletal changes are described in parents but are probably not common. Occurrence in identical twins has been recorded. Since individuals with this disorder are very disabled by the time of early adulthood, family studies of their offspring are not available. Where the chromosomes have been examined they have been found to be normal.

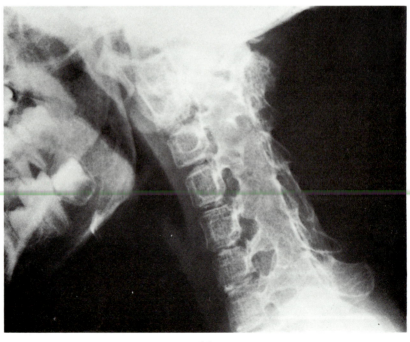

(a)

Figure 10.6. Radiographs of an adolescent boy with MOP who was thought to have Still's disease. (a) Cervical spine, to show fusion

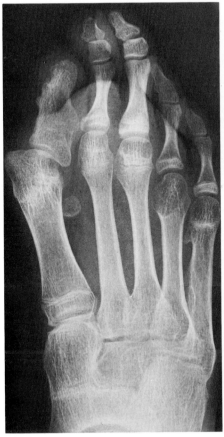
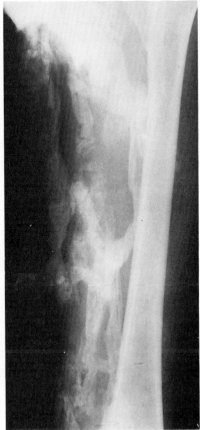

(b)   *(c)*

Figure 10.6. (b) Feet to show digital abnormality. (c) Thigh, to show ossification, which did not become radiologically obvious until the age of 14 years despite progressive stiffness from early childhood

Pathological examination of an established area of ectopic ossification demonstrates that all components of bone are present, including cartilage, with active bone formation and resorption (*Figure 10.7*). There is nevertheless ignorance about the nature of the initial lesion, since although the clinical and macroscopic features suggest that the disorder is primarily one of muscle, i.e. a myositis, there is also evidence that it begins in the connective tissue within muscle. For this reason, McKusick (1972) prefers the term fibrodysplasia ossificans progressiva rather than MOP. We do not know what precipitates any of the episodes of 'myositis', although trauma has been implicated.

Treatment

In myositis ossificans progressiva the main aims of treatment must be to alleviate established disabilities and to prevent progressive loss of mobility. Some help can be given with the former by providing aids for everyday tasks, by

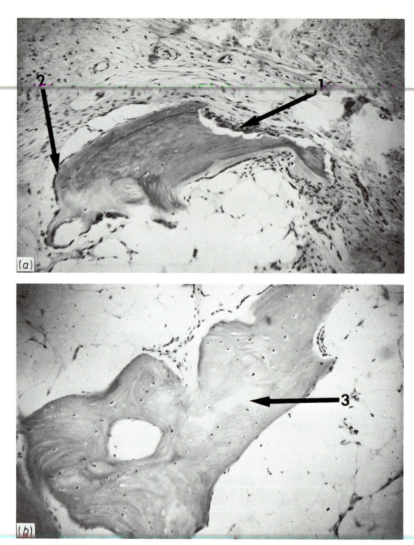

Figure 10.7. The features of ectopic bone in MOP. (a) There is evidence of active bone resorption (1), and osteoblastic bone formation (2); (b) cartilage-like material (3) is also present. (From Smith, 1975)

adapting the environment and also by supplying appropriate wheelchairs. Some measures should be planned in advance, before disability is gross. For instance, since there is often progressive loss of movement of the jaws, any dental work which requires to be done should be done early.

It is natural to consider operations to mobilize joints but surgery on its own has two major drawbacks. First, it is often far more difficult to obtain even a slight increase in mobility by removal of areas of ectopic ossification than the radiographic appearances suggest and, second, the ectopic bone always returns, so that the eventual postoperative state is often worse than that before operation.

DISCUSSION

The prevention of progressive disability may not be possible until we can stop the formation of ectopic bone matrix in muscles. Since we do not know why this occurs measures such as steroids and radiotherapy (for which occasional success has been claimed) remain empirical. Another approach, which is a compromise, accepts that the bone matrix will continue to form but attempts to prevent its mineralization. The chelating agent disodium-EDTA has been used, and also diets low in vitamin D and calcium, but with no convincing success. More recently the diphosphonate EHDP (Chapter 5) has been given to a number of patients, with or without surgery. Initial studies suggested that EHDP could reduce the incidence of myositis (Bassett *et al.*, 1969) and delay recalcification of ectopic bone matrix after its surgical removal (Russell *et al.*, 1972). Subsequent experience in a number of cases from Oxford has not been consistent (Smith, Russell and Woods, 1976). Thus in some patients it was possible to remove ectopic bone and to delay remineralization (assessed radiologically) for periods of at least two years, whereas in others remineralization rapidly occurred. Since there is no accurate way of knowing how much phosphonate was absorbed and accumulated in the tissues, one cannot refute or disprove the suggestion that the variable response was due to variable absorption.

RARE CAUSES OF ECTOPIC MINERALIZATION

There are several of these. Two deserve mention, namely pseudoxanthoma elasticum, an inherited disorder of connective tissue in which deposits of mineral are seen in the tortuous vessels; and calcific tendinitis, in which pain and inflammation around tendons, such as those of the shoulder joint, are associated with ectopic calcification, without detectable abnormality in the blood (Swannell, Underwood and Dixon, 1970).

DISCUSSION

There is no shortage of problems in this area of metabolic disease and although most of the conditions which have been discussed are rare, they can produce much disability. In ectopic mineralization associated with an increase in plasma calcium or phosphate (or both) the biochemical abnormalities provide a basis for treatment which is sometimes successful. It is a mystery why ectopic calcification shows a particular distribution with a particular disorder (compare hyper- and hypoparathyroidism). The appearance of calcinosis in dermatomyositis is quite unpredictable, as is its course; and the occurrence of ectopic ossification after trauma is not satisfactorily explained. It is of considerable interest that γ-carboxyglutamic acid, which may be important in the mineralization of the normal skeleton (Chapter 1), is found in proteins associated with various forms of ectopic calcification. However, it is by the study of MOP that we have perhaps the best chance of finding the cause for ectopic bone formation and various attempts are being made to discover the presumed inherited cellular abnormalities.

REFERENCES

SUMMARY

Soft tissues may mineralize for many reasons. It is important not to miss obvious biochemical abnormalities, since these give clues about treatment. Generalized forms of calcinosis, as with dermatomyositis, often improve spontaneously, but this is not so with ectopic ossification

REFERENCES

Baldursson, H., Evans, E. B., Dodge, W. F. and Jackson, W. T. (1969). Tumoral calcinosis with hyperphosphatemia. A report on a family with incidence in four siblings. *J. Bone Jt Surg.* **51A,** 913–925

Bassett, C. A. L., Donath, A., Macagno, F., Preisig, R., Fleisch, H. and Francis, M. D. (1969). Diphosphonates in the treatment of myositis ossificans. *Lancet* **2,** 845

Bijvoet, O. L. M., Nollen, A. J. G., Slooff, T. J. J. H. and Feith, R. (1974). Effect of a diphosphonate on para-articular ossification after total hip replacement. *Acta Orthop. scand.* **45,** 926–934

Dent, C. E. and Stamp, T. C. B. (1972). Treatment of calcinosis circumscripta with probenecid. *Br. med. J.* **1,** 216–218

Lafferty, F. W., Reynolds, E. S. and Pearson, O. H. (1965). Tumoral calcinosis. A metabolic disease of obscure etiology. *Am. J. Med.* **38,** 105–118

Lutwak, L. (1964). Myositis ossificans progressiva. Mineral, metabolic and radioactive studies of the effects of hormones. *Am. J. Med.* **37,** 269–293

McKusick, V. A. (1972). Heritable disorders of connective tissue. 4th edn. St. Louis: C. V. Mosby Co

Mozaffarian, G., Lafferty, F. W. and Pearson, O. H. (1972). Treatment of tumoral calcinosis with phosphorus deprivation. *Ann. intern. Med.* **77,** 741–745

Nassim, J. R. and Connolly, C. K. (1970). Treatment of calcinosis univeralis with aluminium hydroxide. *Archs Dis. Childh.* **45,** 118–121

Russell, R. G. G. (1975). Diphosphonates and polyphosphates in medicine. *Br. J. Hosp. Med.* **14,** 297–314

Russell, R. G. G., Smith, R., Bishop, M. C., Price, D. A. and Squire, C. M. (1972). Treatment of myositis ossificans progressiva with a diphosphonate. *Lancet* **1,** 10–12

Smith, R. (1975). Myositis Ossificans Progressiva: a review of current problems. *Semins Arthritis Rheum.* **4,** 369–380

Smith, R., Russell, R. G. G. and Woods, C. G. (1976). Myositis ossificans progressiva. Clinical features of eight patients and their response to treatment. *J. Bone Jt Surg.* **58B,** 48–57

Stover, S. L., Niemann, K. M. W., and Miller, J. M. (1976). Disodium etidronate in the prevention of postoperative recurrence of heterotopic ossification in spinal-cord injury patients. *J. Bone Jt Surg.* **58A,** 683–688

Swannell, A. J., Underwood, F. A. and Dixon, A. St. J. (1970). Periarticular calcific deposits mimicking acute arthritis. *Ann. rheum. Dis.* **29,** 380–385

Walker, G. F., Davison, A. M., Peacock, M. and McLachlan, M. S. F. (1977). Tumoral calcinosis: a manifestation of extreme metastatic calcification occurring with 1α-hydroxy-cholecalciferol therapy. *Postgrad. med. J.* **53,** 570–573

Wilber, J. F. and Slatopolsky, E. (1968). Hyperphosphataemia and tumoral calcinosis. *Ann. intern. Med.* **68,** 1044–1049

11

Osteopetroses and Hyperostoses, Hyperphosphatasia and Hypophosphatasia, Fibrous Dysplasia, Fibrogenesis Imperfecta Ossium

INTRODUCTION

Closing chapters in a book of this sort are often reserved for the rare and miscellaneous conditions which do not fit in elsewhere. There is no merit in including diseases merely for the sake of completeness, but the disorders considered here are of interest in their own right. For example, marble bones disease is sufficiently common to be seen (if only as a curiosity) by most orthopaedic surgeons, is inherited and has demonstrable biochemical abnormalities; hypophosphatasia, though very rare, can provide diagnostic difficulties and is recognizable by direct measurement of the enzyme deficiency; fibrous dysplasia comes in different forms and has inexplicable endocrine associations; and finally fibrogenesis imperfecta ossium (FIO), an extreme rarity, appears to have a mineralization disorder secondary to abnormal bone matrix.

OSTEOPETROSES

There are a number of separate conditions characterized by varying combinations of excessive amounts of mineralized bone and defects of bone modelling. These may be called osteopetroses or osteoscleroses, and the names are to some extent

TABLE 11.I

The osteopetroses*

OSTEOSCLEROSES
(Sclerosis predominates, minor changes in bony shape)

Osteopetrosis   (severe recessive and mild dominant)

Pycnodysostosis

CRANIOTUBULAR DYSPLASIAS
(Sclerosis of the cranium and abnormal modelling of long bones)

Metaphyseal dysplasia   (Pyle's disease)

Craniometaphyseal dysplasia

CRANIOTUBULAR HYPEROSTOSES
(Skeletal deformity due to bony overgrowth rather than defective modelling)

Endosteal hyperostosis   (Van Buchem's disease)

Diaphyseal dysplasia   (Camurati–Engelmann Disease)

Idiopathic hyperphosphatasia

MISCELLANEOUS

Osteopathic striata

Melorheostosis

---

* This table is modified from Beighton, Horan and Hamersma (1977) and gives examples of the disorders in each group. Idiopathic hyperphosphatasia is also considered separately (p. 266)

interchangeable. In some the excessive bone predominates and in others the abnormal shape and modelling. This forms the basis of a simple classification by Beighton, Horan and Hamersma (1977) which will be adopted here (Table 11.I). The distinctions are not always clear cut (for instance between Pyle's disease and Van Buchem's disease) and many sub-varieties are described. However the clinical picture is often sufficiently well defined or has certain peculiarities (such as the syndactyly of sclerosteosis) to allow a definite diagnosis to be made. Further classifications, such as those of Rubin (1964) and of the Paris Conference, which are discussed by McKusick (1972), are too complex for the present chapter. The discussion by Gorlin, Spranger and Kozalka (1969) on the genetic craniotubular dysplasias and craniotubular hyperostoses is useful. The skeletal manifestations of congenital or idiopathic hyperphosphatasia lead to its inclusion amongst the craniotubular hyperostoses, but it can also be classified as an inherited disorder with overactivity of alkaline phosphatase (in contrast to the underactivity in hypophosphatasia), a view which is taken here. It is likely that when the biochemical basis for these disorders is further elucidated, considerable reclassification will be necessary. According to Beighton, Horan and Hamersma (1977) the osteopetroses may be divided into the osteoscleroses, the craniotubular dysplasias, the craniotubular hyperostoses and miscellaneous conditions.

Osteoscleroses

Within this group the commonest disorder is that labelled as osteopetrosis, Albers-Schönberg disease or marble bone disease.

Osteopetrosis

This disorder exists in two genetically-distinct forms, a recessively-inherited severe form and a dominantly-inherited mild form (Johnston *et al.*, 1968; Beighton, Horan and Hamersma 1977).

Severe osteopetrosis. The clinical features of this form are consistent. There is severe bony overgrowth from early infancy (the condition has been diagnosed *in utero*) with involvement of the bone marrow and compression of the cranial

TABLE 11.II
Clinical features of severe osteopetrosis. (From Johnston *et al.,* 1968)

Feature	%
Optic atrophy	78
Splenomegaly	62
Hepatomegaly	48
Poor growth	36
Frontal bossing	34
Fractures	28
Loss of hearing	22
Mental retardation	22
Large head	22
Lymphadenopathy	18
Osteomyelitis	18
Genu valgum	16
Facial palsy	10
Pectus deformities	8

nerves. The infant develops anaemia with bruising and bleeding, splenomegaly and hepatomegaly. The anaemia is leuco-erythroblastic and extramedullary erythropoiesis occurs. Growth is poor and optic atrophy common. There is proptosis with prominent frontal bones. The commonest manifestations, according to the survey by Johnston *et al.*, (1968) are shown in Table 11.II. Interesting features are the tendency to pathological fractures and to osteomyelitis. Death usually occurs from haemorrhage or infection early in childhood.

The diagnosis is most often made radiologically. The appearances are remarkable and diagnostic. In addition to increased density of the bones which occurs more or less throughout the skeleton, there are characteristic variations in

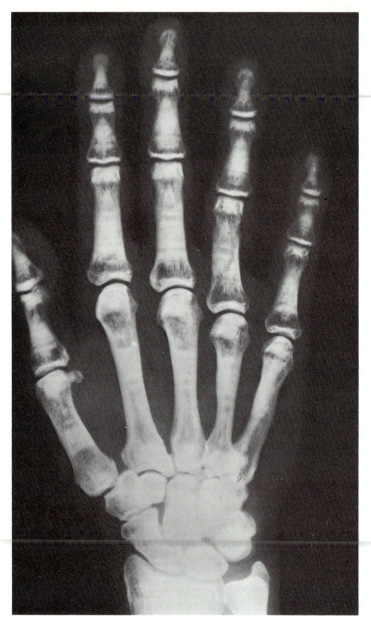

(a)

Figure 11.1. To demonstrate the bony changes in a woman with osteo-petrosis (a, b, c) and her son (d). In this family the disorder is dominantly inherited but has some of the radiographic signs of the recessive severe disorder. (a) The phalanges show the characteristic appearance of 'endobones' (in the phalanges) which have a regular alteration of dense and less dense bone. The 4th and 5th metacarpals are abnormal in shape, and there is evidence of a previous fracture of the 4th proximal phalanx

mineralization (*Figure 11.1*). Thus transverse striations may occur across the metaphyses; concentric layers of increased density may be seen in the developing pelvic bones; and the top and bottom of each vertebral body becomes excessively calcified. In the long bones and phalanges, an appearance of 'bones within bones' (so-called 'endobones') is seen. Modelling of the metaphyses is often defective and produces a club-shaped appearance. In the skull the base is particularly dense, but the vault is also involved.

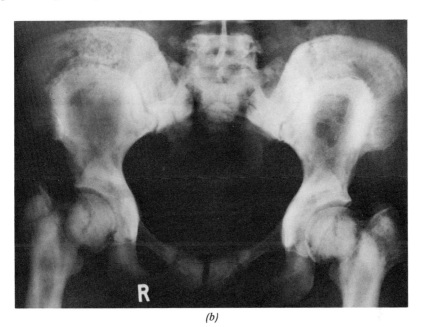

(b)

Figure 11.1. (b) The pelvis and femora show symmetrical areas of alternate dense and less dense bone

The pathology of the bones is of interest. Johnston *et al.* (1968) describe a thick cortex with a marrow cavity the size and shape of that in the foetus. There is little medullary cavity as such, and haemopoiesis occurs in enlarged Haversian canals. There is said to be retardation of resorption and remodelling and a slowing of resorption velocity (Parfitt, 1968). Since there is no reduction in the number of osteoclasts but resorption is apparently ineffective it has been suggested that the bone is abnormal, but Jowsey (1977) states that the osteoclasts are ineffective because of the absence of the ruffled border. Descriptions of the pathological appearances differ, which reflects the heterogeneity of this disorder.

Where they have been investigated the biochemical findings are of considerable interest. The plasma acid phosphatase is often increased, although the alkaline phosphatase is normal; the plasma calcium is usually normal but may be slightly increased. In the severe form of the disease the calcium balance is abnormally positive (for age) with a very low urine calcium. Attempted calcium restriction may produce tetany (Dent, Smellie and Watson, 1965).

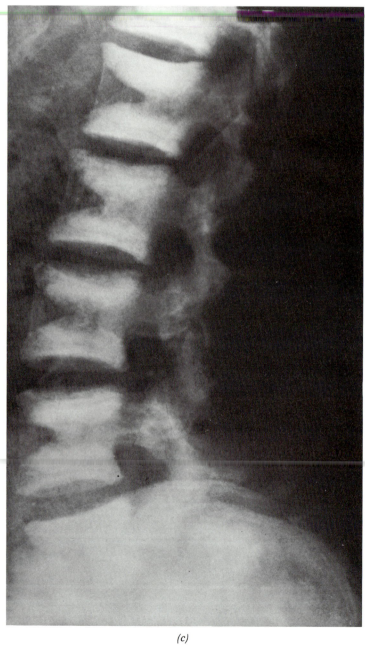

(c)

Figure 11.1. (c) The same alternation is seen in the spine

258

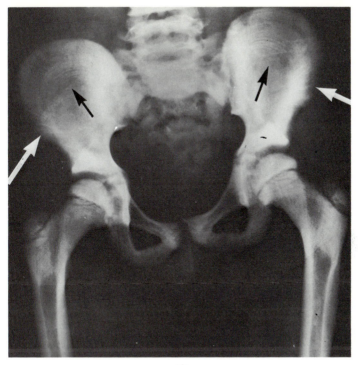

(d)

Figure 11.1. (d) Radiographs of the boy again show variation in density of the bone in the femora and the pelvis. These are fine concentric lines of increased density parallel to the iliac wings (dark arrows)

Advance in our knowledge of the cause and treatment of osteopetrosis has been slow. There are two lines of investigation which are important; that of Dent and his colleagues (1965) on the effects of dietary calcium restriction, and that on experimental transmission of osteopetrosis in mice, particularly by Walker (1975). Dent, Smellie and Watson (1965) described the treatment of a boy with severe osteopetrosis. They reasoned that in this inherited condition the spontaneous variation in density of the bones suggested that the underlying biochemical defect could be affected by the environment, and subsequently demonstrated that periods of severe restriction of dietary calcium coincided with periods of relatively rapid growth and the formation of bone which was less dense and better modelled than usual in this patient. A more complex study by Morrow *et al.,* (1967) was less successful, but Newman (1972) reported encouraging results from calcium depletion.

In the osteopetrotic grey lethal mouse (Walker, 1975 a, b,) and the closely related microphthalmic mouse, bone resorption is defective because there are too few osteoclasts. Normal resorption can be restored by transfusion of splenic or bone-marrow cells from a normal animal after irradiation of the diseased host; conversely the disease can be transmitted to a normal animal. It is suggested that it is the blood-borne precursors of the osteoclasts in the osteopetrotic mouse which are abnormal. In the incisorless rat, bone resorption appears to be defective for another reason, namely inability of the osteoclasts to release lysosomal enzymes. These interesting points are further discussed by Parfitt (1976).

The treatment of the severe form of osteopetrosis has on the whole been unsuccessful. Since the complications (and mortality) of the disease are due to the excessive formation of bone, it is surprising that more attempts have not been made to reduce this, particularly as there is some evidence that this can be done by strict calcium deprivation. Although we do not know whether the overabsorption of calcium is the result or the cause of the bone disease, this need not limit attempts at treatment, especially since osteopetrosis can be produced in some normal animals by a high calcium intake alone. Meanwhile the complications produced by osteopetrosis require attention. Severe anaemia will need blood transfusions and some patients may benefit from corticosteroids or splenectomy.

Recently, Ballet et al., (1977) have described the effect of bone-marrow transplantation in an infant with the severe disorder. The patient had a generalized increase in bone density, with no radiological evidence of a medullary cavity, with hepatosplenomegaly and a leuco-erythroblastic anaemia. The plasma and urine calcium were both low and the kinetic data were said to be typical for osteopetrosis. After marrow transplantation from a 2-year old sister there was continued improvement (for at least 18 months). No further transfusion was required and hepatosplenomegaly disappeared. X-ray showed the appearance of medullary cavities. The authors suggest that in the severe form of the disease bone-marrow transplantation should always be considered, although the human disorder (like that in animals) is likely to be heterogeneous and the results of treatment variable.

Mild osteopetrosis.   This is dominantly inherited, said to be widely distributed and relatively common, and varies (as expected) in its expression. In some subjects abnormal radiographs may be the only sign; in others, fractures, osteomyelitis and involvement of the cranial nerves with facial palsy or deafness occurs. The only consistent biochemical abnormality is the raised plasma acid phosphatase. Metabolic studies by Johnston et al., (1968) on 4 patients with the mild disease showed no consistent abnormality in calcium balance, isotopically-measured calcium turnover, response to injection of parathyroid extract and urinary total hydroxyproline excretion

The radiological appearances in a mother and son with the dominantly-inherited disorder are shown in *Figure 11.1*. The mother, aged 38, is only 4'11" (150 cm) in height. She has eight siblings, all of whom were normal clinically, biochemically and radiologically. She attended a special school, required transfusion during her pregnancies and continues to be anaemic; she has previously sustained fractures. The boy, now aged 11 years, is growing slowly along the 3rd centile and has fractured several bones in his hands and feet. He has a sister who at 8 years is clinically and radiologically normal. Both affected patients have increased plasma levels of acid phosphatase (last recorded results 6.7 and 9.2 K.A. units per 100 ml, respectively – normal less than 3).

The diagnosis of osteopetrosis usually presents no difficulty. It includes a number of the conditions mentioned elsewhere in this chapter; in the mild form in adults, myelofibrosis and fluorosis may also have to be excluded.

Pycnodysostosis

In this rare recessively-inherited disorder, bone sclerosis appears early in child-hood and progresses during growth. There is no disturbance of bone modelling; 'endobones' and bone striations are not seen. The anterior fontanelle remains patent, the cranium bulges but the facial bones are small and the angle of the mandible is obtuse (*see* Beighton, Horan and Hamersma, 1977). The terminal phalanges are short, with the appearance of terminal resorption. The clavicles are hypoplastic.

This is one of several skeletal disorders which Toulouse-Lautrec is reputed to have had (McKusick, 1972).

Craniotubular dysplasias

The disorders which morphologically share various degrees of abnormal cranial sclerosis and tubular remodelling include some in which the clinical features are very mild (such as metaphyseal dysplasia — Pyle's disease) and others which produce grotesque deformity (such as the severe form of craniometaphyseal dysplasia). They are well described by Gorlin, Spranger and Kozalka (1969), and will not be considered in detail here.

Metaphyseal dysplasia

The main clinical feature is knock knee, but muscle weakness, scoliosis and bony fragility are described. The deformity of the knees is due to a striking defect of modelling which is most marked in the distal femora and extends far along the shaft with an 'Erlenmeyer flask' appearance. Similar appearances are seen in the proximal tibia, the proximal humerus and the distal radius and ulna. This disorder is inherited as a recessive.

Craniometaphyseal dysplasia

Two forms of this are described, of which the dominantly-inherited form is relatively common (compared with other conditions in this group) and is less severe than the recessively-inherited form. The changes in the metaphyses are far less than in Pyle's disease and it is the affection of the bones of the skull, face and mandible which give it the disfiguring features.

Paranasal bossing develops during infancy, and the skull and mandible progressively thicken. The seventh and eighth cranial nerves may become compressed, with facial palsy and deafness. In the recessive form all the features are more severe. Optic atrophy and severe distortion of the face and skull may occur.

Craniotubular hyperostoses

In this group of conditions skeletal deformity both of the skull and long bones results from bony overgrowth. The features of Van Buchem's disease (endosteal hyperostosis) and Camurati–Engelmann disease (progressive diaphyseal dysplasia) will be described. This group also includes sclerosteosis, infantile hyperostosis and hyperphosphatasia (*see* p. 266).

Endosteal hyperostosis (Van Buchem's disease)

The descriptions of this disorder (Van Buchem *et al.*, 1962; Jacobs, 1977) differ in details but there is a recognizable clinical picture in which the outstanding feature is the enlargement and sclerosis of the mandible (*Figure 11.2*) which characteristically leads to bilateral facial paralysis and deafness. Other bones

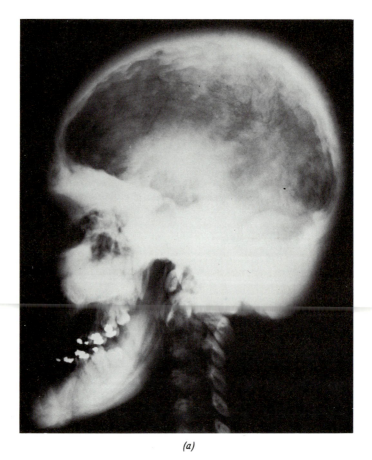

(a)

Figure 11.2. Radiographic changes in an adolescent boy with Van Buchem's disease. (a) The skull shows a massive increase in density, particularly at its base, and the mandible is dense and enlarged

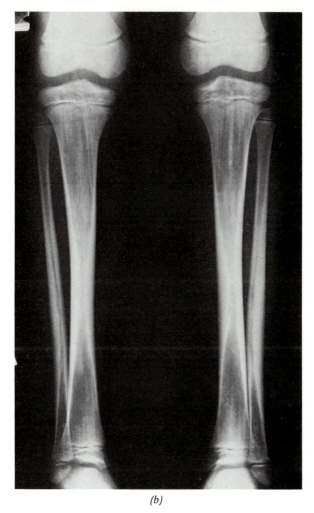

(b)

*Figure 11.2. (b) The increased density of the long bones is due
to endosteal thickening of the diaphyses. Longitudinal striations
are also present*

affected include the calvaria, the base of the skull and the tubular bones where
there is endosteal thickening of the diaphyses. The original descriptions fitted
with an autosomal recessive inheritance but a dominantly-inherited mild disorder
may also occur. Most patients have normal growth and normal intelligence.
Radiographs of a boy with bilateral facial palsy are shown in *Figure 11.2*. In
this patient the long bones also show osteopathia striata. Jacobs (1977) considers
the combination of osteopetrosis and osteopathia striata to be a distinct entity.

Patients with Van Buchem's disease may have a raised alkaline phosphatase,
but allowance has not always been made for the normal increase during
adolescence.

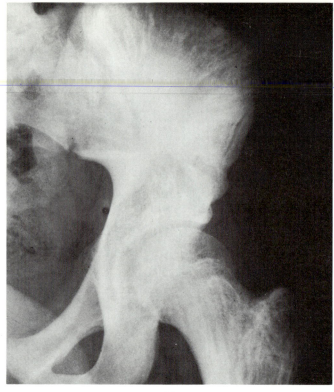

(c)

Figure 11.2. (c) In the pelvis the dense lines radiate out towards the periphery of the ilium. (Compare Figures 11.1b and d)

Diaphyseal dysplasia (Camurati–Engelmann disease)

In this disorder the main feature is diaphyseal hyperostosis. Many variations exist. The condition may be progressive and generalized from childhood, with involvement of practically all the skeleton, or localized with diaphyseal hyperostosis only. The inheritance is autosomal dominant (Sparkes and Graham, 1972).

In the generalized disorder the clinical features are sufficient to establish the diagnosis (Hundley and Wilson, 1973). The bones are painful, walking is difficult and the gait is stiff. The muscles are weak and their mass is strikingly reduced. Bodily disproportion develops, with disproportionately long limbs, which may be partly due to the delayed sexual development. Frontal bossing, proptosis, deafness and blindness can occur. The eventual radiological appearances may be so bizarre (Smith *et al.*, 1977) that it may be necessary to have a series of x-rays during development to make the correct diagnosis. Investigations on these patients have been few. The ESR is characteristically increased. In a recent study of 4 patients (Smith *et al.*, 1977) the alkaline phosphatase and urinary hydroxy-proline were also increased. In one of these patients, the metabolic balance (*see*

(a)

Figure 11.3. (a) The radius and ulna in a female of 19 years with progressive diaphyseal dysplasia (Camurati–Engelmann disease). Thickening started in the diaphysis and extended throughout the length of the bone. (b) The appearance of the lower femur and upper tibia in a severely incapacitated woman of 50 years with the same disorder. (These are Cases 1 and 3 respectively of Smith et al., (1977), where further examples are discussed)

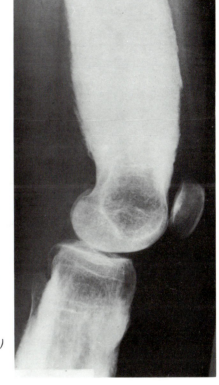

(b)

Figure 2.10) closely resembled that of severe osteopetrosis, with a persistently positive calcium balance and a very low urinary excretion of calcium. This patient, a girl of 19 years, had the clinical and radiographic features (*Figure 11.3*) of the generalized disease, and in addition to the positive calcium balance had hypocalcaemia and hyperphosphataemia which were not explained by the usual causes (Chapter 2). The severity of her illness could be partly due to the fact that her parents were consanguineous.

Treatment is difficult. Relief of the bone pain with corticosteroids has been described (Allen *et al.*, 1970) but successful reduction of excessive bone formation has not been recorded (*see, however,* Fourman and Royer, 1968, Figure 5.4).

Miscellaneous hyperostotic disorders

There are many of these whose genetics and biochemistry await elucidation. They include conditions which are often merely radiological oddities, such as osteopathia striata, osteopoikilosis and melorheostosis (Jowsey, 1977, gives a good example of this last condition). Beighton, Horan and Hamersma (1977) also include Paget's disease in this group, which should at least stimulate ideas about its cause. Certainly in Paget's disease the bones are abnormal, and there is often evidence of a familial tendency (Chapter 5).

HYPERPHOSPHATASIA

The commonly used name for this disorder is derived from the very high level of plasma alkaline phosphatase. This does not imply that the biochemical abnormality is the primary one, any more than the use of the word hypophosphatasia implies a primary defect in this enzyme, although the possibility exists.

Idiopathic (or congenital) hyperphosphatasia is a very rare disease which appears to be recessively inherited. Its main features (Eyring and Eisenberg, 1968; Thomson *et al.,* 1969) are the onset in infancy of skeletal abnormalities with a large head and anterior bowing of the limb bones. The bones are unduly fragile, and the patient is dwarfed. Radiographs particularly show progressive irregular thickening and enlargement of the skull bones and widening of the diaphyses of the long bones. Decreased density of the vertebral bodies and long bones suggests osteoporosis. The characteristic biochemical changes are a sustained rise in plasma alkaline phosphatase and urine hydroxyproline to levels which are only equalled by severe Paget's disease. The bone itself shows extensive fibrosis of the marrow with evidence of considerable cellular overactivity. Accounts differ about the presence of a mosaic pattern of cement lines which are otherwise characteristic of Paget's disease.

Although the clinical picture suggests a single disorder which is inherited as an autosomal recessive, available descriptions emphasize different aspects of the condition. Thus Eyring and Eisenberg (1968) note in their patients a superficial resemblance to osteogenesis imperfecta, with deformed, osteoporotic and fragile bones, blue sclerae, opalescent teeth (which were lost early) and some high-tone deafness; one of the patients of Thompson *et al.,* (1969) developed a large non-malignant tumour containing partly-calcified osteoid and fibrous tissue and showing numerous osteoclasts and osteoblasts; and the patient of Woodhouse *et al.,* (1972), who was less severely affected and said to have juvenile Paget's disease, had a prominent mosaic pattern within iliac-crest bone.

HYPOPHOSPHATASIA

In this condition two biochemical changes can be demonstrated — a reduction in circulating alkaline phosphatase and an increase in urinary phosphoethanolamine. Available evidence suggests that the disorder is recessively inherited, but its severity differs widely according to the time of onset (Fraser, 1957). The cause of the wide variation is not fully understood but is not closely related to the prevailing level of alkaline phosphatase, since patients have occasionally been described with the clinical picture of hypophosphatasia but with normal levels of alkaline phosphatase. Early descriptions such as those of Schlesinger, Luder and Bodian (1955) and Fraser (1957) deal well with the clinical aspects; Rasmussen and Bartter (1978) provide a recent comprehensive review.

Pathology

The central position of alkaline phosphatase in the mineralization process (Chapter 1) suggests why its deficiency might lead to a disturbance of mineralization, although it is not known whether the enzyme defect is primary or whether it is secondary to cellular abnormalities (as for instance in the osteoblast). Since alkaline phosphatase is also a pyrophosphatase its deficiency could lead to an excess of pyrophosphate with inhibition of mineralization. Studies by Russell (1965) did in fact show an increase in urinary pyrophosphate excretion in hypophosphatasia and later (Russell *et al.*, 1971) a similar increase in the plasma. These measurements were made before the demonstration of calcifying vesicles in cartilage, bone and other mineralizing tissues. Since these vesicles have pyrophosphatase activity, any explanation of the effects of hypophosphatasia must take them into account (Rasmussen and Bartter, 1978). Whatever the exact cause the biochemical changes fit well with the observed pathology; this closely resembles rickets and is characterized by defective ossification in cartilage, with widening of the zones of provisional calcification, disruption of the normal column arrangement of cells, and failure of calcification of degenerating cartilage.

Clinical features

Hypophosphatasia is a disorder of very variable severity, and it is most severe when it begins early in life. The condition can therefore be graded according to its time of onset (Fraser, 1957) although these divisions are not sharp;

(1) In the most severe form the disorder may be diagnosed *in utero* or shortly after birth. There is marked demineralization of the skeleton, and the appearance may suggest severe osteogenesis imperfecta. Radiographs show almost complete disappearance of the zone of provisional calcification and the skull appears to have wide clefts between the sutures which in fact represent unmineralized osteoid. Together with the changes in the skeleton there are symptoms of a generalized illness, sometimes with anorexia, irritability, vomiting, dehydration, anaemia and fever. Other described features include blue sclerae, mild skin pigmentation, failure of growth and repeated attacks of cyanosis and pneumonia, and occasional hypercalcaemia.

(2) In the intermediate form, diagnosis is made after 6 months of age. Such patients may have little evidence of generalized illness and come to medical attention because of orthopaedic abnormalities. The bone disturbance is less severe but still widespread, and closely resembles rickets, with bowing of the legs and widening of the ends of the long bones and costochondral junctions. Other characteristic features

include premature loss of deciduous teeth, and premature synostosis of the cranial sutures; exophthalmos and brain damage may follow. There is stunting of growth. Radiologically there are many appearances. The metaphyses are very ragged and there may be translucent clefts extending towards the diaphyses, with irregular calcification. Severe angulation of the long bones, near the larger joints, is described. With synostosis of the skull, the skull bones may develop a copper-beaten appearance due to increased intracranial pressure.

(3) Where the disorder is diagnosed for the first time in adult life (although there may be a past history of rickets) the main feature is undue fragility of the long bones. Hypophosphatasia in the adult is very rare, but it should be considered where, for instance, there are fractures of the femoral shaft of unknown cause. Partial fractures also occur which resemble Looser's zones but are seen on the lateral rather than the medial border of the long bones. There may be a history of 'rickets' in childhood (Bethune and Dent, 1960).

An interesting feature of hypophosphatasia is an increase in chondrocalcinosis (O'Duffy, 1970) presumably related to increased levels of pyrophosphate.

In 1972 Mehes et al., reported the incidence and inheritance of hypophosphatasia in an endogamous Hungarian village. Of 198 schoolchildren between the age of 8 to 14 years, 15 were said to have the hypophosphatasia gene; these all had a plasma alkaline phosphatase of less than 7 K.A. units per 100 ml, increased phosphoethanolamine (PEA) excretion in the urine and clinical signs of the disease, of which the most common was premature loss of deciduous teeth. In 5 parents with symptoms attributed to hypophosphatasia the alkaline phosphatase was persistently normal, although PEA excretion was increased. Another biochemical finding was a reduction in renal-tubular reabsorption of phosphate. This study dealt only with the benign juvenile form of the disease, but was particularly useful in emphasizing the variable expression of hypophosphatasia and the occurrence of the clinical syndrome with normal levels of alkaline phosphatase (as previously described by Pimstone, Eisenberg and Silverman, 1966).

Diagnosis, treatment and genetics

The diagnosis is based on the combination of the characteristic bone disorder, the reduced plasma alkaline phosphatase and presence of excess phosphoethanolamine in the urine, and sometimes affected siblings. The cause of the excess urinary PEA which is demonstrated chromatographically, is unknown. Rasmussen and Bartter (1978) provide a complex diagram which shows the biochemistry of this substance. PEA is said to be increased occasionally in coeliac disease, hypothyroidism and other disorders. Other biochemical features may include hypercalcaemia and hypercalcuria.

The treatment of this condition is difficult. Craniectomy has been performed in severely affected patients with craniostenosis. Since the condition is inherited as a recessive the parents are not usually clinically affected but have a urine phosphoethanolamine excretion which is intermediate between normal levels and those of hypophosphatasia. Thus heterozygotes may be detected but prenatal diagnosis of hypophosphatasia in pregnancies at risk is at present only possible for the severe congenital lethal variants of the disease (Blau *et al.,* 1977). Rudd *et al.,* (1976) report that the amniotic fluid alkaline phosphatase may be normal in an affected pregnancy but that ultrasound, showing defective ossification of the skull, is a consistent finding.

FIBROUS DYSPLASIA

Fibrous dysplasia is a mysterious disease. The localized (monostotic) form is relatively common, and is not associated with other disorders. The generalized (polyostotic) form (often labelled as Albright's syndrome) is characterized by widespread but often unilateral bone disease (*Figures 11.4* and *11.5*), unilateral

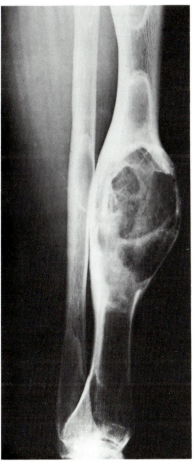 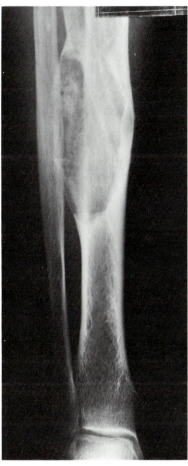

(a)   (b)

Figure 11.4. Comparison of the appearances of (a) fibrous dysplasia; (b) Paget's disease. Both x-rays are from men of about 30 years old. The patient with fibrous dysplasia had other lesions on the same side of the skeleton. In the patient with Paget's disease the lesions rapidly progressed. (Figure 5.2. (a) comes from the same patient)

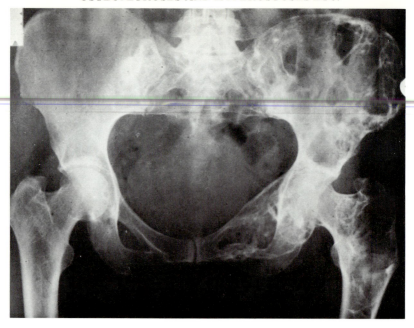

Figure 11.5. The radiological appearance of the pelvis in a woman with polyostotic fibrous dysplasia. The bone lesions are unilateral, and there were similar changes in the thoracic cage and skull on the same side together with cutaneous pigmentation. The bone changes caused spinal cord compression. There was no evidence of endocrine abnormality

pigmentation and (in females) sexual precocity. Thyrotoxicosis, acromegaly, and Cushing's syndrome may also form part of the syndrome for reasons which are quite obscure.

The commonest feature of the monostotic form, which may present at any age, is a fracture. This often occurs in the upper end of the femur, but other bones, the ribs and facial bones, may be involved. Harris, Dudley and Barry (1962) in a review of 50 cases of fibrous dysplasia, of which 37 were polyostotic, and 13 monostotic, considered that they were all varieties of the same disease, although there was no evidence that the monostotic form became polyostotic. Sarcoma developed in 2 patients, but only after irradiation treatment.

The bone changes in the polyostotic form may be extensive and progressive. The 'shepherd's crook' deformity of the femur is characteristic. Spinal cord compression, asymmetry of the skull and leontiasis of the facial bones with disfigurement is described. The cause of the sexual precocity, which is a variable feature, is unknown (see *New England Journal of Medicine*, 1975). An increased end-organ sensitivity to trophic hormones has been postulated (the opposite of pseudohypoparathyroidism). Hypophosphataemic osteomalacia (p. 125) is a rare feature of this condition (Dent and Gertner, 1976). In the generalized form the plasma alkaline phosphatase and the urinary hydroxyproline can be increased.

Treatment can be very difficult, and is not limited to orthopaedic problems. Calcitonin has been used to reduce the overactivity of the abnormal bone, but with little success.

REFERENCES

Diagnosis both of the localized and generalized form should provide no difficulty. However, when Paget's disease occurs in young people it may easily be misdiagnosed as fibrous dysplasia (*Figure 11.4*). Where necessary the distinction can be made by bone histology.

FIBROGENESIS IMPERFECTA OSSIUM (FIO)

The neglect of bone matrix in work on metabolic bone disease was emphasized at the beginning of this book, and it is appropriate to conclude it by considering a rare disorder in which defective mineralization is associated with (and possibly secondary to) an abnormal bone matrix.

In this disorder, fibrogenesis imperfecta ossium (FIO), the characteristic change occurs in the arrangement of bone collagen which is laid down in a disorderly fashion and consequently loses most of its birefringence under polarized light. FIO appears to be acquired, since abnormal collagen is laid down adjacent to normal collagen, and it appears also to be restricted to bone.

The main symptoms, which in recorded cases come on after the age of 50 years (Swan *et al.*, 1976), are severe bone pain and tenderness. Progressive immobility may also occur. Radiologically there are coarse trabeculae throughout the skeleton which may suggest Paget's disease, but the diffuse nature and lack of bony expansion are against this. Ectopic mineralization and ossification are also described (Swan *et al.*, 1976). Biochemically the only consistent change is an increase in plasma alkaline phosphatase.

Examination of the bone shows that the trabeculae are increased in width though reduced in number. Histologically these are wide osteoid seams with lack of birefringence. Good illustrations are given by Swan *et al.* (1976).

There is no established treatment, although the patient of Baker *et al.* (1966) responded temporarily to vitamin D_2 and DHT. The cause is quite unknown, but Henneman, Pak and Bartter (1972) reported that the collagen of bone (but not of skin) was excessively soluble in neutral salt solution, in acid, and in EDTA.

CONCLUSION

This book has concentrated on conditions which affect the skeleton and result from or produce biochemical abnormalities. It has been difficult to know what to exclude, especially since scientific progress continually uncovers new metabolic changes in disease. In the future further causes of the inherited disorders of the skeleton will doubtless continue to be identified, not only in present-day curiosities but in more common conditions, such as achondroplasia or neuro-fibromatosis. For the moment these diseases have been excluded, but room has been left for stop-press information (Chapter 12). The skeleton may look inactive but work upon it is not.

REFERENCES

Allen, D. T., Saunders, A. M., Northway, W. H., Williams, G. P. and Schafer, I. A. (1970). Corticosteroids in the treatment of Engelmann's disease. Progressive Diaphyseal Dysplasia. *Pediatrics* **46**, 523–531

Baker, S. L., Dent, C. E., Friedman, M. and Watson, L. (1966). Fibrogenesis imperfecta ossium. *J. Bone Jt Surg.* **48B**, 804–825

REFERENCES

Ballet, J. J., Griscelli, C., Coutris, C., Milhaud, G. and Maroteaux, P. (1977); Bone-marrow transplantation in osteopetrosis. *Lancet* **2**, 1137

Beighton, P., Horan, F. and Hamersma, H. (1977). A review of the osteopetroses. *Postgrad. med. J.* **53**, 507–515

Bethune, J. and Dent, C. E. (1960). Hypophosphatasia in an adult. *Am. J. Med.* **28**, 615–622

Blau, K., Hoar, D. I., Rattenbury J. M. and Rudd, N. L. (1977). Prenatal diagnosis of hypophosphatasia. *Lancet* **2**, 1139

Dent, C. E. and Gertner, J. M. (1976). Hypophosphataemic osteomalacia in fibrous dysplasia. *Q. Jl Med.* **45**, 411–420

Dent, C. E., Smellie, J. M. and Watson, L. (1965). Studies in osteopetrosis. *Archs Dis. Childh.* **40**, 7–15

Eyring, E. J. and Eisenberg, E. (1968). Congenital hyperphosphatasia. A clinical, pathological and biochemical study of two cases. *J. Bone Jt Surg.* **50A**, 1099–1117

Fourman, P. and Royer, P. (1968). *Calcium Metabolism and the Bone.* 2nd edn. Oxford: Blackwell Scientific Publications

Fraser, D. (1957). Hypophosphatasia. *Am. J. Med.* **22**, 730–746

Gorlin, R. J., Spranger, J. and Kozalka, M. F. (1969). Genetic craniotubular bone dysplasias and hyperostoses. A critical analysis. In First Conference on the Clinical Delineation of Birth Defects. *Birth defects. Original Article Series* Vol. V. No. 4. April, 1969. The National Foundation – March of Dimes. pp. 79–95

Harris, W. H., Dudley, H. R. and Barry, R. J. (1962). The natural history of fibrous dysplasia. An orthopaedic, pathological and roentgenographic study. *J. Bone Jt Surg.* **44A**, 207–233

Henneman, D. H., Pak, C. Y. C. and Bartter, F. C. (1972). Collagen composition, solubility and biosynthesis in fibrogenesis imperfecta ossium. In *Clinical Aspects of Metabolic Bone Disease. Excerpta med.* p. 469

Hundley, J. D. and Wilson, F. C. (1973). Progressive diaphyseal dysplasia. Review of the literature and report of seven cases in one family. *J. Bone Jt Surg.* **55A**, 461–474

Jacobs, P. (1977). Van Buchem's disease. *Postgrad. Med. J.* **53**, 497–505

Jowsey, J. (1977). *Metabolic Bone Disease.* Vol. I. London: W. B. Saunders Co

Johnston, C. C., Lavy, N., Lord, T., Vellios, F., Merritt, A. D. and Deiss, W. P. (1968). Osteopetrosis. A clinical, genetic, metabolic and morphologic study of the dominantly-inherited benign form. *Medicine* **47**, 149–167

Mehes, K., Klujber, L., Lassu, G. and Kajtar, D. (1972). Hypophosphatasia: screening and family investigations in an endogamous Hungarian village. *Clin. Genet.* **3**, 60–66

McKusick, V. A. (1972). *Heritable Disorders of Connective Tissue.* St. Louis: C. V. Mosby Co

Morrow, G., Barness, L. A., Fost, A. and Rasmussen, H. (1967). Calcium mobilization in osteopetrosis. *Am. J. Dis. Child.* **114**, 161–168

New England Journal of Medicine (1975). MGH Case Records, **292**, 199–203

Newman, C. L. (1972). Osteopetrosis. *Proc. R. Soc. Med.* **65**, 729–730

O'Duffy, J. D. (1970). Hypophosphatasia associated with calcium pyrophosphate dihydrate deposits in cartilage. *Arthritis Rheum.* **13**, 381–387

Parfitt, A. M. (1976). The actions of parathyroid hormone on bone: relation to bone remodelling and turnover, calcium homeostasis and metabolic bone disease. Part IV. *Metabolism* **25**, 1157–1188

Flinstone, D. D., Disenborg, E. and Silverman, S. (1966). Hypophosphatasia. Genetic and dental studies. *Ann. intern. Med.* **65**, 722

Rasmussen, H. and Bartter, F. C. (1978). Hypophosphatasia. In *The Metabolic Basis of Inherited Disease.* Eds J. B. Stanbury, J. B. Wyngaarden and D. S. Fredrickson. 4th edn. pp. 1340–1349. New York: McGraw-Hill

Rudd, N. L., Miskin, M., Hoar, D. I., Benzie, R. and Doran, T. A. (1976). Prenatal diagnosis of hypophosphatasia. *New Engl. J. Med.* **295**, 146–148

Rubin, P. (1964). *Dynamic Classification of bone Dysplasias.* Chicago: Year Book Medical Publishers

Russell, R. G. G. (1965). Excretion of inorganic pyrophosphate in hypophosphatasia. *Lancet* **2**, 461–464

Russell, R. G. G., Bisaz, S., Donath, A., Morgan, D. B. and Fleisch, H. (1971). Inorganic pyrophosphate in plasma in normal persons and in patients with hypophosphatasia, osteogenesis imperfecta and other disorders of bone. *J. clin. Invest.* **50**, 961–969

Schlesinger, B., Luder, J. and Bodian, M. (1955). Rickets with alkaline phosphatase deficiency: an osteoblastic dysplasia. *Archs Dis. Childh.* **30**, 265–276

REFERENCES

Smith, R., Walton, R. J., Corner, B. D. and Gordon, I. R. S. (1977). Clinical and biochemical studies in Engelmann's disease (progressive diaphyseal dysplasia). *Q. Jl Med.* **46**, 273–294

Sparkes, R. S. and Graham, C. B. (1972). Camurati–Engelmann disease. Genetics and clinical manifestations with a review of the literature. *J. med. Genet.* **9**, 73–85

Swan, C. H. J., Shah, K., Brewer, D. B. and Cooke, W. T. (1976). Fibrogenesis Imperfecta Ossium. *Q. Jl Med.* **45**, 233–253

Thompson, R. C., Gaull, G. E., Horwitz, S. J. and Schenk, R. K. (1969). Hereditary hyperphosphatasia. *Am. J. Med.* **47**, 209–219

Van Buchem, F. S. P., Hadders, N. N., Hansen, J. F. and Woldring, M. G. (1962). Hyperostosis corticalis generalizata. Report of seven cases. *Am. J. Med.* **33**, 387–397

Walker, D. G. (1975a). Bone resorption restored in osteopetrotic mice by transplantation of normal bone marrow and spleen cells. *Science N.Y.* **190**, 784–785

Walker, D. G. (1975b). Spleen cells transmit osteopetrosis in mice. *Science N.Y.* **190**, 785–787

Woodhouse, N. J. Y., Fisher, M. T., Sigurdsson, G., Joplin, G. F. and MacIntyre, I. (1972). Paget's disease in a 5-year-old: acute response to human calcitonin. *Br. med. J.* **4**, 267–269

12

Recent Advances

This brief chapter is intended to keep the reader up-to-date and give some guide to the growing points in biochemical disorders of the skeleton. Like the preceding chapters it does not aim to be comprehensive in so short a space, and since relevant new articles on currently popular and controversial subjects appear in virtually every issue of the weekly journals, selection is necessary. Examples of such subjects include the significance of low-plasma calcitonin in women; the therapeutic place of the 1α-hydroxylated metabolites of vitamin D; the role of mithramycin in Paget's disease; and the identification of the different types of osteogenesis imperfecta. The recent work will be considered in chapter order.

There are two useful accounts of collagen (*Lancet* 1978a; Pinnell, 1978) which supplement the information in Chapter 1. One of the most striking aspects of collagen is its almost universal distribution. No longer should it be regarded merely as a protein of fibrous tissues, since it clearly has other important functions in basement membranes, in platelet aggregation and in tissues, such as muscle, during their development. Although most biochemical advance has come from a study of the rare but well-defined inherited disorders, the major application of such advance may eventually be in acquired diseases of collagen. In addition to the defects in collagen described in Chapters 1 and 8, a preliminary report that Type I collagen is reduced in Marfan's syndrome is referred to (*Lancet,* 1978a). Where molecular changes have been identified it is not always easy to relate them to the clinical features. Pinnell (1978) deals with clinically relevant features of collagen synthesis and structure and particularly with the various disorders classified under the Ehlers—Danlos syndrome. Although the disorders in this syndrome do not primarily affect the skeleton, it would be foolish to disregard them on this account since they can tell us a lot about collagen.

Calcitonin (Chapter 1) may now be measured within the physiological range in plasma, although the recorded 'normal' values differ according to the method used. In recent studies Hillyard, Stevenson and MacIntyre (1978) confirmed that levels of immunoreactive calcitonin are low in women compared with men, and are increased by oral oestrogen — progestagen pills, and by pregnancy.

Cundy *et al.* (1978) compared the levels in women who had fractures of the femoral neck with an unfractured control group and found no significant difference. Thus it seems possible that the reported low levels of calcitonin may have some relevance to postmenopausal osteoporosis, if not to fracture. Hillyard *et al.* (1978) emphasize the importance of family-screening in familial medullary carcinoma of the thyroid. They point out the importance of considering this diagnosis in a male with a thyroid mass.

In osteoporosis (Chapter 3) two useful articles have appeared. An editorial (*British Medical Journal,* 1978a) emphasizes the difference between prevention of osteoporosis and the treatment of the established condition, and gives useful references to recent work. In controlled trials, oestrogens prevent the 1—4 per cent bone loss per year which otherwise occurs in the first menopausal decade; the bone loss in those taking calcium supplements alone was intermediate between the two groups. In older women who do not absorb calcium effectively, vitamin D or its metabolites can increase this absorption but may lead to hypercalcaemia or hypercalcuria. The possible deleterious effects of oestrogens are not yet defined. Likewise beneficial effects of human PTH (1—34), growth hormone, or fluoride plus vitamin D on established osteoporosis are not yet established. Perhaps the most striking effect of this potentially-preventable disorder is that which refers to the fractured neck of the femur; it is said that the annual (1978) cost of this will exceed 200 million US dollars for Europe and North America. Lindsay *et al.* (1978) have followed up their work on the prevention (and reversal) of post-oöphorectomy osteoporosis by oestrogen treatment, and show that when such patients stop oestrogens the loss of bone is rapid. The result of this is that oöphorectomized patients who had had four year's treatment with oestrogens, followed by four years without oestrogens, had a bone-mineral content which was not significantly different from a group treated for eight years with a placebo. This finding, which suggests that in order to maintain bone mass oestrogen treatment must be continued, has important implications.

Articles on osteomalacia and vitamin D (Chapter 4) still exceed (in number and volume) those on any other subject in this book. The causes of the high incidence of osteomalacia in biliary cirrhosis (*Lancet,* 1978d), after bowel resection (*British Medical Journal,* 1978b), and after jejuno-ileal bypass for obesity (Compston *et al.,* 1978), are still being discussed but impaired metabolism of vitamin D (in biliary cirrhosis) and malabsorption of this vitamin (after bowel resection and bypass) must be important. The description of osteomalacia in β thalassemia (Gratwick *et al.,* 1978) suggests a new field for investigation. The importance of bone biopsy in the diagnosis of osteomalacia has again been emphasized (*Lancet,* 1978d). Kooh *et al.,* (1977) have resurrected the idea that rickets may occasionally be due to calcium deficiency and Brooks *et al.,* (1978) have discussed a form of vitamin-D-dependent rickets with possible end-organ resistance to $1,25(OH)_2D$.

The causes of renal and dialysis osteodystrophy continue to be investigated. Animal work has shown the importance of a raised plasma phosphate in stimulating parathyroid overactivity, presumably via hypocalcaemia, and control of hyperphosphataemia is therefore considered important to prevent bone resorption (*British Medical Journal,* 1978c). Animal experiments also show that after $\frac{7}{8}$ nephrectomy, survival can be prolonged by phosphate restriction (Ibels *et al.,* 1978) but there is no direct evidence that phosphorus restriction in man will slow the decline of renal function in chronic renal failure. The problem is not a simple one and the effects of injudicious phosphate depletion are many (*Lancet*, 1978e). Attempts are still being made to find the cause of dialysis bone disease; excessive aluminium is the latest in a list of possible culprits (Ward *et al.*, 1978).

Work and writing on the clinical uses of the hydroxylated metabolite of vitamin D continues unabated, and is the subject of two editorials (*Lancet*, 1978b; *British Medical Journal*, 1978d). Winterborn *et al.* (1978) suggest that 1α-(OH)D (and $1,25(OH)_2D$) may have a deleterious effect on renal function. Kanis, Cundy and Naik (1978), who have analysed their available data on this, point out the difficulty in detecting such an effect, especially since the rate at which renal function deteriorates tends to accelerate as renal failure advances. Thus it is very difficult to separate the renal effects of the 1α-hydroxy derivatives of vitamin D from the natural history of renal failure. The striking effect of $1,25(OH)_2D$ on renal osteodystrophy and growth in children has been described by Chesney *et al.* (1978), supplementing similar work from Oxford (Chapter 4). Adequate alkali treatment of children with renal-tubular acidosis also appears to increase growth rate and it has been suggested that this is due to correction of the block between $25(OH)D$ and $1,25(OH)_2D$ which acidosis produces (McSherry and Morris, 1978). The bone disease of renal failure is likely to be multifactorial and $25(OH)D$ deficiency should not yet be disregarded. Bordier *et al.* (1978), who gave either $1\alpha(OH)D_3$ or $1,25(OH)_2D_3$ to 6 patients (3 on dialysis) for only 3 weeks suggested that factors in addition to the 1α-hydroxylated metabolites may control bone formation and mineralization in such patients, but such brief studies are necessarily of limited interest. Of other such metabolites, $24,25(OH)_2D$ has recently come under investigation (Kanis *et al.*, 1978); current evidence suggests that it is biologically active in man and may have a different action from $1,25(OH)_2D$, causing calcium retention (presumably in the skeleton) without hypercalcaemia. When the dust has settled it may turn out that the main therapeutic use of such 1α-hydroxylated metabolites is in the prevention and treatment of renal osteodystrophy, and the main advantages over the native vitamin is their speed of action, short half-life and ease of control. Since dihydroxytachysterol (DHT) has some of these features, it should not be neglected.

The epidemiology and treatment of Paget's disease (Chapter 5) have received further attention. Thus, in men, the incidence of Paget's disease in British migrants to Western Australia is higher than in native Australians but lower than in U.K. residents (Gardner, Guyer and Barker, 1978). This suggests that the migrants either develop the disorder before they go or take it with them in a latent form, and that environment is also important. Controversy continues about the treatment of Paget's disease; it is considered by some that the pain of uncomplicated Paget's disease can nearly always be controlled by analgesics, by others that calcitonin is effective in virtually homeopathic doses (Milhaud, 1978), and by others that mithramycin is either very useful or too toxic for use. Hamdy (1978) writes that the use of specific antipagetic drugs as a first-line treatment for pain is unjustified. There are few who would disagree with this, and most patients who are subsequently treated with calcitonin, mithramycin or diphosphonates are those who have failed to respond to simpler measures. Evans and McIntyre (1978) point out that although small doses of calcitonin may improve pain, larger doses are necessary to stop bone resorption. Calcitonin is often very effective for pain which does not respond to analgesics. The exact role of phosphonates is not yet established. A recent review (*Lancet*, 1978c) and studies by Canfield *et al.* (1977) and Khairi (1977) not referred to in Chapter 5 suggest that the best effective daily dose of EHDP which avoids possible side-effects is 5 mg per kg body weight, for not more than six months. In diagnosis isolated reports refer to the coexistence of Paget's disease and hyperparathyroidism. They serve to emphasize the important clinical point that hypercalcaemia in a patient with Paget's disease should not too readily be attributed to immobilization, especially if the plasma phosphate is low.

The cause of primary hyperparathyroidism (Chapter 6) is still obscure, but there are now several reports in which parathyroid adenomas have occurred in patients who have had previous irradiation of the neck and head (Schachner and Hall, 1978). In the patient with hypercalcaemia and malignant disease without metastases it is worthwhile considering that the hypercalcaemia is due to co-existent hyperparathyroidism; diagnosis of this may be aided by measurement of nephrogenous cAMP (Drezner and Lebovitz, 1978). Hypercalcaemia continues to be reported in a variety of non-malignant and malignant conditions which even include Burkitt's lymphoma (Spiegel *et al.,* 1978). Freitag *et al.* (1978) provide evidence that the considerable increase in carboxy-terminal iPTH in renal failure (*see Figure 6.5*) is due to impaired metabolism of PTH by the damaged kidney. Studies continue on the cause and treatment of pseudo-hypoparathyroidism (PHP). There is some evidence that in this disorder the adenylate cyclate enzyme may be abnormal. Treatment with $1,25(OH)_2D$ is highly effective but hypercalcaemia has been reported associated with an un-explained increase in circulating $1,25(OH)_2D$ after cessation of treatment.

Following the report by Levin, Salinas and Jorgenson, (1978) on the division of dominantly-inherited osteogenesis imperfecta according to dental character-istics, the classification of this disorder (Chapter 7) has come under further scrutiny by Sillence and Rimoin (1978). These authors refer to a comprehensive study of osteogenesis imperfecta in Victoria, Australia, which suggests that at least four different syndromes exist, in some of which the sclerae are normal-coloured or become so with age.

Recent investigation of homocystinuria (Chapter 8) supports its heterogeneity. Thus the study of the cystathionine β-synthase activity in fibroblasts from 14 synthase-deficient patients suggests at least three general classes of mutants (Fowler *et al.,* 1978).

The possibility that collagen may be abnormal in idiopathic scoliosis (Chapter 8) has been supported by the data of Bushell, Ghosh and Taylor (1978), who measured the extractability by pepsin of collagen from different areas of inter-vertebral discs in a normal and scoliotic subject after death. The results were compatible with a cross-linking defect in collagen from the scoliotic subject, but there were many other possibilities.

One of the hallmarks of hypophosphatasia (Chapter 11) is an increased excretion of phosphoethanolamine (PEA) and it is important to know how much this is increased in other conditions. Licata and his colleagues (1978) have studied this; they find that although the PEA/creatinine ratio in the urine alters with the time of day, age, bone disease and protein intake, high levels occur only in hypophosphatasia.

Finally, one should note important chapters relevant to the skeleton which have appeared in the 4th edition of *The Metabolic Basis of Inherited Disease* (Stanbury, Wyngaarden and Fredrickson, 1978) including reviews on collagen and its disorders; on the mucopolysaccharidoses; on homocystinuria; on pseudohypoparathyroidism; and on vitamin-D-resistant and vitamin-D-dependent rickets.

The last two subjects are reviewed by Rasmussen and Anast (1978). While the cause of the very rare vitamin-D-dependent rickets appears to be fairly clearly due to 1α-hydroxylase deficiency, the basic cause of hypophosphataemic rickets remains unknown. This review demonstrates two important points on which it is appropriate to conclude; first, how it is possible for physicians to make an apparently simple disease exceedingly complex (see Preface); and second, that the best management of vitamin-D-resistant rickets and many other biochemical disorders of bone comes from close co-operation between physicians and orthopaedic surgeons.

REFERENCES

Bordier, Ph., Zingraff, J., Gueris, J., Jungers, P., Marie, P., Pechet, M. and Rasmussen, H. (1978). The effect of 1α(OH)D$_3$ and 1α,25(OH)$_2$D$_3$ on the bone in patients with renal osteodystrophy. *Am. J. Med.* **64**, 101–107

British Medical Journal (1978a). Treatment of osteoporosis (Editorial). **1**, 1303–1304

British Medical Journal (1978b). Short bowels (Editorial). **1**, 737–738

British Medical Journal (1978c). Hyperparathyroidism in renal failure (Editorial). **1**, 390–391

British Medical Journal (1978d). Clinical use of 1α-hydroxy-vitamin D$_3$ (Editorial). **1**, 1571–1572

Brooks, M. H., Bell, N. H., Love, L., Stern, P. H., Orfei, E., Queener, S. F., Hamstra, A. J. and deLuca, H. F. (1978). Vitamin-D-dependent rickets Type II. *New Engl. J. Med.* **298**, 996–999

Bushell, G. R., Ghosh, P. and Taylor, T. K. F. (1978). Collagen defect in idiopathic scoliosis. *Lancet* **2**, 94–95

Canfield, R., Rosner, W., Skinner, J., McWhorter, J., Resnick, L., Feldman, F., Kammerman, S., Ryan, K., Kunigonis, M. and Bohne, W. (1977). Diphosphonate therapy of Paget's disease of bone. *J. Clin. Endocr. Metabol.* **44**, 96–106

Chesney, R. W., Moorthy, A. V., Eisman, J. A., Jax, D. K., Mazess, R. B. and de Luca, H. F. (1978). Increased growth after long-term oral 1α25 vitamin D$_3$ in childhood renal osteodystrophy. *New Engl. J. Med.* **298**, 238–242

Compston, J. E., Horton, L. W. L., Laker, M. F., Ayers, A. B., Woodhead, J. S., Bull, H. J., Gazet, J. C. and Pilkington, T. R. E. (1978). Bone disease after jejuno-ileal bypass for obesity. *Lancet* **2**, 1–4

Cundy, T., Heynen, G., Ackroyd, C., Kissin, M., Kirby, R. and Kanis, J. A. (1978). Plasma calcitonin in women. *Lancet* **2**, 159

Drezner, M. K. and Lebovitz, H. E. (1978). Primary hyperparathyroidism in paraneoplastic hypercalcaemia. *Lancet* **1**, 1004–1006

Evans, I. M. A. and McIntyre, I. (1978). Treatment of Paget's disease. *Lancet* **2**, 213

Fowler, B., Kraus, J., Packman, S. and Rosenberg, L. E. (1978). Homocystinuria. Evidence for three distinct classes of cystathionine β-synthase mutants in cultured fibroblasts. *J. clin. Invest.* **61**, 645–653

Freitag, J., Martin, K. J., Hruska, K. A., Anderson, C., Conrades, M., Ladenson, J., Klahr, S. and Slatopolsky, E. (1978). Impaired parathyroid hormone metabolism in patients with chronic renal failure. *New Engl. J. Med.* **298**, 29–32

Gardner, M. J., Guyer, P. B. and Barker, D. J. P. (1978). Radiological prevalence of Paget's disease of bone in British migrants to Australia. *Br. med. J.* **1**, 1655–1657

Gratwick, G. M., Bullough, P. G., Bohne, W. O., Markenson, A. L. and Peterson, C. M. (1978). Thalassemia osteoarthropathy. *Ann. intern. Med.* **88**, 494–501

Hamdy, R. (1978). Mithramycin in Paget's disease. *Lancet* **1**, 1267

Hillyard, C. J., Stevenson, J. C. and MacIntyre, I. (1978). Relative deficiency of plasma-calcitonin in normal women. *Lancet* **1**, 961–962

Hillyard, C. J., Evans, I. M. A., Hill, P. A. and Taylor, S. (1978). Familial medullary thyroid carcinoma. *Lancet* **1**, 1009–1011

REFERENCES

Ibels, L. S., Alfrey, A. C., Haut, L. and Huffer, W. E. (1978). Preservation of function in experimental renal disease by dietary restriction of phosphate. *New Engl. J. Med.* **298**, 122–126

Kanis, J. A., Cundy, T., and Naik, R. (1978). 1α-hydroxy derivatives of vitamin D and renal function. *Lancet* **2**, 316–317

Kanis, J. A., Cundy, T., Bartlett, M., Smith, R., Heynen, G., Warner, G. T. and Russell, R. G. G. (1978). Is 24,25-dihydroxycholecalciferol a calcium-regulating hormone in man? *Br. med. J.* **1**, 1382–1386

Khairi, M. R. A., Altman, R. D., DeRosa, G. P., Zimmermann, J., Schenk, R. K. and Johnston, C. C. (1977). Sodium Etidronate in the treatment of Paget's disease of bone. *Ann. intern. Med.* **87**, 656–663

Kooh, S. W., Fraser, D., Reilly, B. J., Hamilton, J. R., Gall, D. G. and Bell, L. (1977). Rickets due to calcium deficiency. *New Engl. J. Med.* **297**, 1264–1266

Lancet (1978a). Collagen in health and disease (Editorial). **1**, 1077–1079

Lancet (1978b). 'One-alpha' (Editorial). **1**, 973–974

Lancet (1978c). Ten years treatment for Paget's disease (Editorial). **1**, 914–915

Lancet (1978d). Bone biopsy and vitamin D in primary biliary cirrhosis (Editorial). **1**, 1138

Lancet (1978e). Hyperphosphataemia and renal function (Editorial). **1**, 753–754

Levin, L. S., Salinas, C. F. and Jorgenson, R. J. (1978). Classification of osteogenesis imperfecta by dental characteristics. *Lancet* **1**, 332–333

Licata, A. A., Radfar, N., Bartter, F. C. and Bou, E. (1978). The urinary excretion of phosphoethanolamine in diseases other than hypophosphatasia. *Am. J. Med.* **64**, 133–138

Lindsay, R., Hart, D. M., MacLean, A., Clark, A. C., Kraszewski, A. and Garwood, J. (1978). Bone response to termination of oestrogen treatment. *Lancet* **1**, 1325–1327

McSherry, E. and Morris, R. C. (1978). Attainment and maintenance of normal stature with alkali therapy in infants and children with classic renal-tubular acidosis. *J. clin. Invest.* **61**, 509–527

Milhaud, G. (1978). Low-dose calcitonin in Paget's disease. *Lancet* **1**, 1153

Pinnell, S. R. (1978). Disorders of collagen. In *The Metabolic Basis of Inherited Disease.* Eds J. B. Stanbury, J. B. Wyngaarden and D. S. Fredrickson. 4th edn. pp. 1366–1394. New York: McGraw-Hill

Rasmussen, H. and Anast, C. (1978). Familial hypophosphatemic (vitamin-D-resistent) rickets and vitamin-D-dependent rickets. In *The Metabolic Basis of Inherited Disease.* Eds J. B. Stanbury, J. B. Wyngaarden and D. S. Fredrickson. 4th edn. pp. 1537–1562. New York: McGraw-Hill

Schachner, S. H. and Hall, A. (1978). Parathyroid adenoma and previous head and neck irradiation. *Ann. intern. Med.* **88**, 804

Sillence, D. O. and Rimoin, D. L. (1978). Classification of osteogenesis imperfecta. *Lancet* **1**, 1041–2

Spiegel, A., Greene, M., MacGrath, I. Balow, J., Marx, S. and Aurbach, G. D. (1978). Hypercalcaemia with suppressed parathyroid hormone in Burkitt's lymphoma. *Am. J. Med.* **64**, 691–695

Stanbury, J. B., Wyngaarden, J. B. and Fredrickson, D. S. (Eds). (1978). *The Metabolic Basis of Inherited Disease.* 4th edn. New York: McGraw-Hill

Ward, M. K., Feest, T. G., Ellis, H. A., Parkinson, I. S., Kerr, D. N. S., Herrington, J. and Goode, G. L. (1978). Osteomalacic dialysis osteodystrophy: evidence for a water-borne aetiological agent, probably aluminium. *Lancet* **1**, 841–845

Winterborn, M. H., Pace, P. J., Heath, D. A. and White, R. H. R. (1978). Impairment of renal function in patients on 1α-hydroxycholecalciferol. *Lancet* **2**, 150–151

Glossary of Abbreviations

ATP	Adenosine triphosphate
ATP'ase	Adenosine triphosphatase
cAMP	Cyclic AMP, $3'5'$ adenosine monophosphate
C1q	A collagen-like component of complement
CT	Calcitonin, thyrocalcitonin
25(OH)D	25-hydroxycholecalciferol*
1α(OH)D	1α-hydroxycholecalciferol
$1,25(OH)_2D$	1,25-dihydroxycholecalciferol, a metabolite of vitamin D produced by the kidney
ECF	Extracellular fluid
EHDP	Disodium etidronate, disodium ethane-1-hydroxy-1, 1-diphosphonate
GFR	Glomerular filtration rate
α_2HS glycoprotein	A circulating glycoprotein incorporated into bone
MPS	Mucopolysaccharidosis
OAF	Osteoclast activating factor
Pase	Alkaline phosphatase (AP)
PGE_2	Prostaglandin E_2
PHP	Pseudohypoparathyroidism
PPHP	Pseudopseudohypoparathyroidism

* *See* footnote to p. 24

GLOSSARY OF ABBREVIATIONS

PIP	Proline iminopeptidase
PTH	Parathyroid hormone
THP	Total urinary hydroxyproline
TmP	Maximum rate of tubular reabsorption (for phosphate)
^{99m}Tc	Microaggregated technetium 99
Vitamin B_6	Pyridoxine, pyridoxal phosphate

Index

289

INDEX

Teeth, (*cont.*)
  metabolic bone disease, in, 45
  osteogenesis imperfecta, in, 184, 186, 195
Tendinitis, calcific, 251
Testosterone, osteoporosis caused by deficiency of, 31
Tetany,
  metabolic bone disease, in, 48
  rickets, in, 48
Thrombosis in homocystinuria, 211, 212, 213, 215
Thyrotoxicosis, 270
  bone loss in, 79
  osteoporosis in, 91
Thyroxine, 30
  growth, effect on, 38
Tissue damage, calcification secondary to, 241
Tissue paper scars, 212
Toe, shortened big, 247
Trauma, ossification following, 246, 251
Trident hand, 46
Trousseau's sign, 48
Tumoral calcinosis, 240
Tumour rickets, 126
Turner's syndrome,
  bone loss in, 79, 80
  dwarfism in, 68
  neck webbing in, 46
  osteoporosis in, 91

Urine,
  alkaptonuria, in, 220, 222
  calcium in, 53, 58
    hyperparathyroidism, in, 166
    osteomalacia and rickets, in, 107
    osteoporosis, in, 80
  creatinine in, 54
  examination in metabolic bone disease, 53
  homocystine in, 215
  hydroxyproline in, 54
    collagen breakdown and, 15
  keratan sulphate in, 234, 236
  mucopolysaccharides in, 237
  phosphate in, 56

Van Buchem's disease, 254, 262
Vertebral body,
  collapse in osteoporosis, 86
  homocystinuria, in, 215
  Hurler syndrome, in, 231
  hyperphosphatasia, in, 266
  osteomalacia, in, 109
  osteoporosis, in, 74, 79, 80, 83, 89, 91
  Paget's disease, in, 140, 142
Vitamin B_6-responsive homocystinuria, 211, 213, 216, 217
Vitamin D, 2, 23–27
  action on target organs, 25, 26
  coeliac disease and, 103
  deficiency,
    osteomalacia, in, 95
    sex differences, 117
  dietary sources, 24
  hypoparathyroidism, in, 175
  metabolism in, 21, 102
    renal failure, in, 103
  muscle, effect on, 102
  osteomalacia and, 275
  osteoporosis, in, 83
  overdosage, 49
  parathyroidectomy and, 172
  synthesis of, 24
  treatment of rickets, in, 124
Vitamin-D-dependent rickets, 127, 278
Vitamin D metabolites, 24, 52, 276
  action on target organs, 25
  nomenclature of, 24*n*
  renal osteodystrophy, in, 58
  renal osteomalacia, in, 122
  rickets and, 102
Vitamin-D-resistant rickets, 98, 123
  hypophosphataemic, 98
  phosphate levels in, 51

Weightlessness, skeletal effects of, 78
Wilson's disease, 42
  eyes in, 44
  osteomalacia and rickets in, 101
Wormian bones, 192

Zollinger–Ellison syndrome, 28, 172